IMO
표준해사통신영어

IMO Standard Marine Communication Phrases

박진수·김현종 공저

해광출판사

머 리 말

인류의 역사가 시작된 이래, 인간의 음성은 상호간의 통신에 가장 중요한 수단이 되어 왔지만, 그 사용범위가 너무도 제한되어 있는 것이 사실이다. 따라서 해상에서는 통신수단으로서 신호용 기류, 발광신호기, 음향신호기, 확성장치, 무선통신기 등을 사용하게 되었으며, 근래에는 해상통신의 대부분이 무선통신에 의하여 이루어지고 있다. 또, 1999년 2월부터 GMDSS가 전면적으로 시행됨에 따라 해상 무선통신은 종래의 지상파를 이용한 MORSE 통신으로부터 INMARSAT 등 인공위성이 포함된 VOICE, TELEX 통신 위주로 바뀌게 되어 그 신뢰성, 간편성, 사용범위 등이 비약적으로 향상되었으며, 대부분 항해사가 이 통신을 담당하게 되었다.

그러나 아직도 해결해야할 큰 문제가 있으니, 이는 사용 언어가 서로 다를 경우에 발생하는 의사 소통상의 문제이다. 사실, 이러한 문제는 항공 분야에 있어서도 마찬가지로서, 항공기 운항의 고속성 등으로 인하여 통신용 언어를 통일해야 할 필요성은 해상보다도 더 긴요하여 오래 전부터 항공 통신용 언어인 “Air English"가 사용되어 오고 있었다.

해상 분야에서도 해상통신용어의 필요성을 인식하게 되어, IMO총회는 1977년 11월에 표준해사항해용어집(SMNV)을 채택하여 해상에서 사용하도록 하였고, 1985년 5월에 이에 대한 일부의 개정이 이루어져 사용되어오다가, 1992년 이래 IMO의 해사안전위원회(MSC) 주도하에 더 포괄적이고 표준화한 안전언어를 개발하여왔다. 1997년의 제68차 MSC 회의에서는 표준해사통신용어(SMCP)초안을 채택하였고, 이 초안은 1년여에 걸친 국제적 시험 사용 뒤에, 각국의 의견을 수렴하여 2001년 11월에 IMO총회의 결의서 A.918(22)로 그 최종안이 채택되었다.

이 교재는 IMO가 채택한 SMCP를 바탕으로, 해상에서 의사소통에 사용되는 언어와 전문용어를 표준화하여 영한 대역본으로 출간함으로서, 해사교육기관의 학생은 물론, 승선 중인 해기사들의 의사소통에 도움을 주어 항해안전에 기여하고자 한다. (2006년 가을)

또한 이 증보개정판에서는 VHF 사용에 관한 국제지침서 등을 참조하여 올바른 VHF 무선전화 사용방법을 추가하였으며, 필요한 곳에 “예문”을 삽입하여, 독자들의 이해를 돕고자 하였다. 특히 지난 30여년간 해사영어 교육훈련에 힘써오면서, IMO SMCP 개발을 주도해온 Prof. Peter Trenkner에게 감사의 인사를 드리며, 또한 이 책의 편집에 참여해준 “Maritime English Honors Club” 학생 여러분의 노고에 감사드린다. (2016년 가을)

STANDARD MARINE COMMUNICATION PHRASES

THE ASSEMBLY,

RECALLING Article 15(j) of the Convention on the International Mari-time Organization concerning the functions of the Assembly in relation to regulations and guidelines concerning maritime safety,

RECALLING ALSO resolution A.380(X) by which it adopted the Standard Marine Navigational Vocabulary,

RECALLING FURTHER the provisions of regulation V/14.4 of the International Convention for the Safety of Life at Sea, 1974, requiring that on all ships to which chapter I thereof applies, English shall be used on the bridge as the working language for bridge-to-bridge and bridge-to-shore safety communications as well as for communications between the pilot and bridge watchkeeping personnel unless those directly involved in the communications speak a common language other than English;

RECOGNIZING that the standardization of language and terminology used in such communications would assist the safe operation of ships and contribute to greater safety of navigation,

RECOGNIZING ALSO the wide use of the English language for inter- national navigational communications and the need to assist maritime training institutions to meet the objectives of safe operations of ships and enhanced navigational safety through, *inter alia*, the standardization of language and terminology used,

HAVING CONSIDERED the recommendations of the Maritime Safety Committee at its sixty-eighth and seventy-fourth sessions,

1. ADOPTS the IMO Standard Maritime Communication Phrases set out in Annex 1 to the present resolution;

표준 해사 통신 용어(SMCP)

총회는,

해사 안전 분야의 규정 및 지침에 관련된 총회의 기능에 관한 IMO협약 제 15조(j)항을 상기하고,

표준 해사 항해 영어를 채택하였던 결의서 A.380(x)를 또한 상기하며,

SOLAS협약 제 1장이 적용되는 모든 선박에 있어서, 선교와 선교 사이, 선교와 육상 사이의 안전 통신 뿐만 아니라, 교신에 직접 참여하는 사람들이 영어 이외의 공통 언어를 사용하지 않는 한, 선교 당직 요원과 도선사 사이의 의사소통에 있어서도 업무용 언어로서 영어가 선교에서 사용되어야 한다고 요구하는 1974년의 국제 해상 인명 안전 협약의 규정 V/14.4의 조항들을 또한 상기하며,

그러한 의사소통에 사용되는 언어와 전문용어를 표준화함으로써 선박의 안전 운항에 도움을 주고 항해의 안전에 더 큰 공헌을 하리라는 점을 인식하고,

국제 항해상의 통신에 영어가 광범위하게 사용되는 점과, 선박의 안전 운항이라는 목표를 달성할 수 있도록 해사 훈련 기관을 도울 필요성 및 특히, 사용되는 언어와 전문 용어를 표준화함으로써 항해의 안전이 증진된다는 점을 또한 인식하며,

해사 안전 위원회 제 68차 및 74차 회의의 권고 사항을 검토하여

1. 이 결의서의 부속서 1에 기술한 표준 해사 통신 용어를 채택한다.

2. AUTHORIZES the Maritime Safety Committee to keep the IMO Standard Marine Communication Phrases under review and to amend them when necessary in accordance with the procedure set out in Annex 2 to the present resolution;

3. RECOMMENDS Governments to give the IMO Standard Marine Commu-nication Phrases a wide circulation to all prospective users and all maritime education authorities, in order to support compliance with the standards of competence as required by table A-II/1 of the STCW Code.

4. REVOKES resolution A.380 (X).[1]

---

1) A.380(10) Standard Marine Navigational Vocabulary

2. 해사 안전 위원회로 하여금 필요한 경우, 본 결의서의 부속서 2에 기술된 절차에 따라 표준 해사 통신 용어를 검토하고 개정할 수 있는 권한을 부여한다.

3. 각국 정부가 이 표준 해사 통신 용어를 예상되는 모든 사용자와 모든 해사 교육 기관에 널리 보급하여 STCW 협약의 표 A-II/1에 의하여 요구되는 해기 능력 기준을 충족시키는 데 도움을 주도록 권고한다.

4. 결의서 A.380(x)는 폐기한다.

FOREWORD

As navigational and safety communications from ship to shore and vice versa, ship to ship, and on board ships must be precise, simple and unambiguous, so as to avoid confusion and error, there is a need to standardize the language used. This is of particular importance in the light of the increasing number of internationally trading vessels with crews speaking many different languages since problems of communication may cause misunderstandings leading to dangers to the vessel, the people on board and the environment.

In 1973 the Maritime Safety Committee agreed at its twenty-seventh session that, where language difficulties arise, a common language should be used for navigational purposes and that language should be English. In consequence the Standard Marine Navigational Vocabulary(SMNV) was developed, adopted in 1977 and amended in 1985.

In 1992 the Maritime Safety Committee, at its sixtieth session, instruct-ed the Sub-Committee on Safety of Navigation to develop a more com-prehensive standardized safety language than the SMNV, 1985, taking into account the changing conditions in modern seafaring and covering all major safety-related verbal communications.

At its sixty-eighth session in 1997 the Maritime Safety Committee adopted the Draft Standard Marine Communication Phrases(SMCP) developed by the Sub-Committee on Safety of Navigation. The draft SMCP, following international trials, was amended at the forty-sixth session of this Sub-Committee and final consideration given at the Maritime Safety Committee at its seventy-fourth session in the light of remarks received by the Organization. The SMCP was adopted by the Assembly in 22nd session as resolution A.918(22).

Under the International Convention on Standards of Training, Certification and Watchkeeping for Seafarers, 1978, as revised 1995, the ability to understand and use the SMCP is required for the certification of officers in charge of a navigational watch on ships of 500 gross tonnage or more.

서 문

혼동과 실수를 피하기 위하여 선박과 육상 간, 선박과 선박 간 및 선내에서 사용되는 항해, 안전 통신은 정확, 간결, 명백하여야 하므로 사용되는 언어를 표준화 할 필요가 있다. 의사 소통상의 문제가 오해를 불러와서, 선박, 승선 중인 사람, 환경에 위험을 초래할 수가 있기 때문에, 여러 가지 서로 다른 언어를 사용하는 선원을 태우고 국제 무역에 종사하는 선박의 숫자가 증가하고 있음을 고려하면, 이 점은 특히 중요하다.

1973년 해사안전위원회(MSC)는 제 27차 회의에서 언어상의 장애가 일어나는 경우, 항해상의 목적을 위하여 공통 언어가 사용되어야 하고 그 언어는 영어가 되어야 한다고 합의하였다. 그 결과 1977년에 표준해사항해용어집(SMNV)이 개발되었고, 1985년에 개정되었다.

1992년에 MSC는 제 60차 회의에서 항해안전전문위원회(NAV)로 하여금, 변화하는 현대의 해상 환경을 고려하고 안전에 관련된 모든 주요 음성통신 사항을 망라하여 1985년의 SMNV보다 더 포괄적이고 표준화한 안전 언어를 개발할 것을 지시하였다.

MSC는 1997년의 제 68차 회의에서 NAV가 개발한 표준 해사 통신 용어(SMCP) 초안을 채택하였다.
SMCP 초안은 국제적인 시험 사용을 거쳐 제46차 NAV 회의에서 수정되었으며, MSC는 IMO에 접수된 비평들을 고려하여 제 74차 회의에서 이를 최종적으로 검토하였다.
총회는 2001년에 결의서 A.918(22)로 SMCP를 채택하였다.

1995년에 개정된 1978년의 선원의 훈련, 자격 증명 및 당직 근무 기준에 관한 국제 협약(STCW)에 의하여, 총톤수 500톤 이상의 선박에서 항해 당직을 담당할 사관의 자격 증명에 SMCP를 이해하고 사용할 수 있는 능력이 요구된다.

STANDARD MARINE COMMUNICATION PHRASES
CONTENTS

표준 해사 통신 용어(SMCP)
목 차

IMO STANDARD MARINE COMMUNICATION PHRASES : PART A

표준 해사 통신 용어: A편

Appendix to AI - External Communication Phrases

AI 편 - 외부 통신 용어에 대한 부록

STANDARD MARINE COMMUNICATION PHRASES : PART B

표준 해사 통신 용어: B편

INTRODUCTION

1 Position of the SMCP in maritime practice

The Standard Marine Communication Phrases(SMCP) has been compiled:

☞ to assist in the greater safety of navigation and of the conduct of the ship,

☞ to standardize the language used in communication for navigation at sea, in port-approaches, in waterways, harbours and on board vessels with multilingual crews, and

☞ to assist maritime training institutions in meeting the objectives mentioned above.

These phrases are not intended to supplant or contradict the International Regulations for Preventing Collisions at Sea, 1972 or special local rules or recommendations made by IMO concerning ships' routeing, neither are they intended to supersede the International Code of Signals, and when applied in ship's external communication this has to be done in strict compliance with the relevant radiotelephone procedures as set out in the ITU Radio Regulations.[1] Furthermore, the SMCP, as a collection of individual phrases, should not be regarded as any kind of technical manual providing operational instructions.

The SMCP meets the requirements of the STCW Convention, 1978, as revised, and of the SOLAS Convention, 1974, as revised, regarding verbal communications; moreover, the phrases cover the relevant communication safety aspects laid down in these Conventions.[2)]

Use of the SMCP should be made as often as possible in preference to other wording of similar meaning; as a minimum requirement users should adhere as closely as possible to their wording in relevant situations. In this way they are intended to become an acceptable safety language, using English for the verbal interchange of intelligence among individuals of all maritime nations on the many and varied

일 러 두 기

1. 해사실무에 있어서 표준 해사 통신 용어(SMCP)의 위치

표준 해사 통신 용어(SMCP)는 다음의 목적으로 편찬되었다.

☞ 선박의 항해와 항행관제의 안정성을 보다 높이는 데 기여하고,

☞ 해상, 항구의 부근, 항로, 항내와 여러 나라 말을 사용하는 선원이 승무하는 선내에서 항행상의 통신에 사용되는 언어를 표준화 하며,

☞ 해사 훈련 기관들로 하여금 상기의 목적을 달성하도록 돕기 위하여.

이 용어집은 1972년의 국제해상충돌예방규칙, 선박의 항로 지정에 관한 특정 해역 규정 또는 IMO의 권고를 대체하거나 부정하는 것은 아니다.
또한 국제신호서를 대체하고자 하는 것이 아니며, 선박의 외부 통신에 이를 활용할 경우에는 ITU의 전파 법규에 규정된 관련 무선 전화 절차를 엄격하게 지켜 실시하여야 한다.
또한, 개개의 용어들의 집합체인 SMCP가 운용상의 지침을 제공하는 기술적 교범으로 간주되어서도 안 된다.

SMCP는 개정된 1978 STCW 협약과 개정된 1974 SOLAS 협약의 음성 언어 통신에 관한 요구 사항을 충족하며 이들 협약에 의하여 설정된 통신 관련 안전 요건에 부합된다.

관련된 상황 하에서 사용자들의 용어를 이에 최대한으로 일치시켜야 하는 최소한의 요구 사항으로서, 유사한 의미를 갖는 다른 용어에 우선하여, 가능한 한 자주 이 SMCP를 사용하여야 한다.
이는 현대의 해상 조건하에서 더욱 더 분명해지고 있는 바로서, 모든 해양국가의 개인 간에 이루어지는 음성에 의한 정보 교환에 있어서 정확한 의미와 해석이 의문시되는 많은 다양한 경우에 대하여, 이와 같이 함으로써 SMCP는 영어를 사용한 적절한 안전 용어가 될 수 있을 것이다.

occasions when precise meanings and translations are in doubt, increasingly evident under modern conditions at sea.

The accompanying CD/Cassette is designed to familiarize users with the pronunciation of the phrases.

2 Organization of the SMCP

The SMCP is divided into External Communication Phrases and On board Communication Phrases as far as its application is concerned, and into PART A and PART B as to its status within the framework of the STCW, 1978, as revised.[3)]

PART A covers phrases applicable in external communications and which may thus be regarded as the replacement of the Standard Marine Navigational Vocabulary, 1985, which is requested to be used and understood by the STCW Code, 1995, Table A-II/I. This part was enriched by essential phrases concerning shiphandling and safety of navigation to be used in on-board communications, particularly when the Pilot is on the bridge, as required by Regulation 14(4), Chapter V, SOLAS 1974, as revised.

PART B calls attention to other on-board standard safety-related phrases which, supplementary to PART A, may also be regarded useful for Maritime English instruction.

3 Position of the SMCP in Maritime Education and Training

The SMCP does not intend to provide a comprehensive Maritime English syllabus which is expected to cover a far wider range of language skills to be achieved in the fields of vocabulary, grammar, discourse abilities, etc., than the SMCP could ever manage. However, PART A in particular, should be an indispensable part of any curriculum which is designed to meet the corresponding requirements of the STCW Convention, 1978, as revised. In addition, PART B offers a rich choice of situations covered by phrases well suited to meet the

부수되는 CD/Cassette는 사용자로 하여금 본 용어집의 올바른 발음을 익히는데 도움을 주도록 기획되었다.

2. SMCP의 구성

SMCP는 그 적용과 관련하여 외부 통신 용어와 선내 통신 용어로 나누어져 있으며, 개정된 1978년 STCW협약의 틀에 맞추어서 그 지위에 따라 A편과 B편으로 나누어졌다.

A편은 외부 통신에 적용되는 용어들을 포함하고 있으며, 1985년의 표준 해사 항해 용어를 대체하는 것으로 간주되고, 1995년 STCW협약의 표 A-Ⅱ/1에 의하여 이의 활용과 이해가 요구된다.
이 부분에는 개정된 1974년 SOLAS 제5장 규칙 14(4)에 의하여 요구되는 바대로, 특히 도선사가 선교에 승무할 경우에, 선내 통신에 사용되어야 할 선박 조종과 항해 안전에 관계되는 필수적인 용어가 첨가되었다.

B편은 여타의 선내 안전 관련 표준 용어에 주의를 환기시키고 있으며, A편에 부가하여 해사 영어의 교육에 유용한 것으로 간주된다.

3. 해사 교육과 훈련에 있어서 SMCP의 위치

SMCP는, 어휘, 문법, 회화 능력 등의 분야에서 SMCP가 감당할 수 있는 것보다 훨씬 더 광범위한 언어 기능을 달성할 것으로 기대되는 해사 영어의 포괄적인 강의 요목을 제공하고자 하지는 않는다. 그러나 특히 A편은, 개정된 1978년 STCW협약의 해당 요구사항에 부응하도록 입안된 어떤 교과과정에 있어서도 **필수 불가결한 부분**이다.
추가하여 B편은, 다양한 상황에 대하여 개정된 1978년 STCW협약의 통신 요구 사항에 잘 부합하도록 망라된 많은 용어를 제공하며, 그것들은 항해자들에게 만족스러울 것으로 기대 된다.

communication requirements of the STCW Convention, 1978, as revised, which are implicitly expected to be satisfied by mariners.[4)]

The SMCP should be taught and learnt selectively, according to the users` specific needs rather than completely. The respective instruction should be based on practice in the maritime environment and be implemented through appropriate modern language teaching methods.

4 Basic communicative features

The SMCP builds on a basic knowledge of the English language. It was drafted on purpose in a simplified version of Maritime English to reduce grammatical, lexical and idiomatic varieties to a tolerable minimum, using standardized structures for the sake of its functional aspects, i.e. diminishing misunderstanding in safety related verbal communications, thereby endeavouring to reflect present Maritime English language usage on board vessels and in ship-to-shore /ship-to-ship communications.

This means, in phrases offered for use in emergency and other situations developing under considerable pressure of time or psychological stress as wells as in navigational warnings, a block language was applied which sparingly uses, or frequently omits, the function words *the, a/an, is/are* as done in seafaring practice. Users, however, may be flexible in this respect.

Further communicative features may be summarized as follows:

- avoiding synonyms
- avoiding contracted forms
- providing fully worded answers to "yes/no"-questions and basic alternative answers to sentence questions
- providing *one* phrase for *one* event, and
- structuring the corresponding phrases after the principle: *identical invariable plus variable.*

SMCP는 전부를 배우기보다는 사용자의 특정한 필요에 따라서 선택적으로 가르치고 배워야 한다. 각각의 교육은 해사 환경 하에서의 실습에 기반을 두어야 하며 적절한 현대적인 언어 교수법에 의하여 실시되어야 한다.

4. 기본적인 통신상의 특징

SMCP는 영어의 기초적인 지식 위에서 구성되었다.

SMCP는 안전 관련 음성 통신에 있어서 오해를 감소시키고자 하는 그 기능적인 측면을 위하여, 의도적으로 문법, 어휘, 관용구상의 다양성을 허용 최소한으로 축소하도록 표준화된 구문을 사용하여 해사영어의 간단한 형식으로 입안되었으며, 그리함으로서 선내 및 선박과 육상/선박과 선박 사이의 통신에 있어서 사용되는 오늘날의 해사영어의 용법을 반영하도록 노력하였다.

이는 비상상황, 시간적으로 상당히 촉박한 상황 또는 심리적으로 압박받는 상황 및 항행 경보 하에서 사용하도록 제시된 용어에 있어서는, 해상에서의 관례에 따라, 언어의 집단이 사용되었으며 이에는 *the, a/an, is/are* 같은 기능어가 드물게 쓰이거나 자주 생략되었음을 의미한다. 그러나 이러한 측면에 있어서는 사용자에게 융통성이 있다.

그 이상의 통신상의 특징은 다음과 같이 요약된다:

- ☞ 동의어의 회피
- ☞ 축약된 형태의 회피
- ☞ "예/아니오" 형 질문에 대하여 생략되지 않은 전체 용어를 사용한 응답을 제공하였으며, 문장형 질문에 대하여는 기본적인 선택의 여지를 주는 응답을 제공하였다
- ☞ 하나의 사건에 대하여 하나의 어귀를 제공하였으며,
- ☞ 원형에 뒤이어서 대응하는 어귀를 구성함 :
  즉 동일한 불변 사항에 뒤이어서 가변사항을 둠

5 Typographical conventions

( ) **brackets** indicate that the part of the message enclosed within the brackets may be added where relevant;

/ **oblique strokes** indicate that the items on either side of the stroke are alternatives;

… **dots** indicate that the relevant information is to be filled in where the dots occur;

(*italic letters*) indicate the kind of information requested;

~ **tildes** precede possible words or phrases which can be use after/in association with the given standard phrase.

5. 인쇄용 관례 기호

( )　　**괄호**는 그 안에 있는 통신문을 적절한 경우 첨가할 수 있음을 나타낸다;

/　　**사선**은 그 앞 뒤에 있는 항목 중 하나를 선택하여 사용할 수 있음 을 나타낸다;

…　　**점선**은 그 부분에 적절한 정보를 삽입할 수 있음을 나타낸 다;

*(괄호 안의 이탤릭체 글씨)*는 요구되는 정보의 종류를 표시한다;

~　　**(반복을 피하기 위한 생략 기호)** 주어진 표준 어귀 다음에/와 관련 지어서 사용될 수 있는 단어나 어귀 앞에 쓰였다.

GENERAL

1 Procedure

When it is necessary to indicate that the SMCP are to be used, the following message may be sent:

"Please use Standard Marine Communication Phrases."

"I will use Standard Marine Communication Phrases."

2 Spelling

2.1 Spelling of letters

When spelling is necessary, only the following spelling table should be used:[5]

Letter	Code	Letter	Code
A	Alfa	N	November
B	Bravo	O	Oscar
C	Charlie	P	Papa
D	Delta	Q	Quebec
E	Echo	R	Romeo
F	Foxtrot	S	Sierra
G	Golf	T	Tango
H	Hotel	U	Uniform
I	India	V	Victor
J	Juliet	W	Whisky
K	Kilo	X	X-ray
L	Lima	Y	Yankee
M	Mike	Z	Zulu

2.2 Spelling of digits and numbers

A few digits and numbers have a **modified** pronunciation compared to general English:[6]

총 칙

1. 절차

이 표준 해사 통신 용어를 사용하여 교신하고자 한다는 것을 나타낼 필요가 있는 경우에는 다음과 같은 통신문을 송신한다:

"표준 해사 통신 용어를 사용하시오"

"표준 해사 통신 용어를 사용하겠습니다"

2. 철자

2.1 글자에 대한 철자

철자를 사용할 필요가 있을 때, 다음의 철자표 만을 사용해야한다:

Letter	Code	Letter	Code
A	Alfa	N	November
B	Bravo	O	Oscar
C	Charlie	P	Papa
D	Delta	Q	Quebec
E	Echo	R	Romeo
F	Foxtrot	S	Sierra
G	Golf	T	Tango
H	Hotel	U	Uniform
I	India	V	Victor
J	Juliet	W	Whisky
K	Kilo	X	X-ray
L	Lima	Y	Yankee
M	Mike	Z	Zulu

2.2 숫자와 수에 대한 철자

일부의 숫자와 수에 대한 발음이 일반적인 영어와 다르게 변경되었다.

Number	Spelling	Pronunciation
0	zero	ZEERO
1	one	WUN
2	two	TOO
3	three	TREE
4	four	FOWER
5	five	FIVE
6	six	SIX
7	seven	SEVEN
8	eight	AIT
9	nine	NINER
1000	thousand	TOUSAND

3 Message Markers

In shore-to-ship and ship-to-shore communication or radio communication in general, the following eight Message Markers may be used (also see **"Application of Message Markers"** given in PART AI/6 "Vessel Traffic service (VTS) Standard Phrases"):

(i) Instruction
(ii) Advice
(iii) Warning
(iv) Information
(v) Question
(vi) Answer
(vii) Request
(viii) Intention

4 Responses

4.1 When the answer to a question is in the affirmative, say:

"Yes, …. " - followed by the appropriate phrase in full.

Number	Spelling	Pronunciation
0	zero	ZEERO
1	one	WUN
2	two	TOO
3	three	TREE
4	four	FOWER
5	five	FIVE
6	six	SIX
7	seven	SEVEN
8	eight	AIT
9	nine	NINER
1000	thousand	TOUSAND

3. 통보 부호

선박과 육상국 간의 통신이나 일반 무선 통신에 있어서 다음 여덟 가지의 통보 부호가 사용될 수 있다. (AI/6 편의 "선박 통항 관제 업무의 표준 용어" 란에 기술된 "통보 부호의 적용"을 참조할 것):

(i) 지시 (Instruction)
(ii) 권고 (Advice)
(iii) 경고 (Warning)
(iv) 정보 (Information)
(v) 질문 (Question)
(vi) 응답 (Answer)
(vii) 요구 (Request)
(viii) 의도 (Intention)

4. 응답

4.1 질문에 대해서 긍정적인 답을 하는 경우에는:

"Yes, …" 라고 말한 다음, 어귀를 생략하지 않고 부가한다.

4.2 When the answer to a question is in the negative, say:

"No, …" - followed by the appropriate phrase in full.

4.3 When the information requested is not immediately available, say:

"Stand by" - followed by the time interval within which the information will be available.

4.4 When the information requested cannot be obtained, say:

"No information."

4.5 When an INSTRUCTION (e.g. by a VTS-Station, Naval vessel or other fully authorized personnel) or an ADVICE is given, respond if in the affirmative:

"I will/can … " - followed by the instruction or advice in full;

if in the negative, respond:

"I will not/cannot … " - followed by the instruction or advice in full.

Example: "ADVICE. Do not overtake the vessel North of you."

Respond: "I will not overtake the vessel North of me."

4.6 Responses to orders and answers to questions of special importance both in external and onboard communication are given in wording in the phrases concerned.

5 Distress, urgency and safety signals[7)]

5.1 MAYDAY to be used to announce a distress message

4.2 질문에 대해서 부정적인 답을 하는 경우에는:

"No, …" 라고 말한 다음, 어귀를 생략하지 않고 부가한다.

4.3 요구된 정보를 즉시 통보해 줄 수 없는 경우에는:

"Stand by" 라고 말한 다음, 요구된 정보를 통보해 줄 수 있을 때까지 기다려야 할 시간을 알려준다.

4.4 요구된 정보를 입수할 수가 없을 경우에는:

"No information" 이라고 말한다.

4.5 지시(즉, VTS국, 해군함정 또는 여타의 전권이 주어진 사람에 의한)나 권고가 통보되었고, 이에 대한 긍정적인 응답을 하고자 할 경우에는:

"I will/can …" 라고 하고, 뒤에 지시나 권고의 내용을 생략하지 않고 반복한다; 그리고,

이에 대한 부정적인 응답을 하고자 할 경우에는:

"I will not/cannot …" 라고 하고, 뒤에 지시나 권고의 내용을 생략하지 않고 반복한다.

보기: "권고. 귀선의 북방에 있는 선박을 추월하지 마시오"

응답: "본선은 본선의 북방에 있는 선박을 추월하지 않겠음"

4.6 외부 통신이나 선내 통신에 있어서, 명령에 대한 응답이나 특히 중요한 내용의 질문에 대한 대답을 할 경우에는, 관련된 어귀 내에 포함되어 있는 용어를 사용한다.

5. 조난, 긴급 및 안전 신호

5.1 메이데이 조난 통보를 알리기 위해서 사용해야 한다.

5.2 PAN-PAN to be used to announce an urgency message

5.3 SĖCURITĖ to be used to announce a safety message

6 Standard organizational phrases

6.1 "How do you read (me)?"

6.1.1 **"I read you …"**

bad/one	with signal strength one(i.e. barely perceptible)
poor/two	with signal strength two(i.e. weak)
fair/three	with signal strength three(i.e. fairly good)
good/four	with signal strength four(i.e. good)
excellent/five	with signal strength five(i.e. very good)

6.2 When it is advisable to remain on a VHF Channel/ frequency say:

"Stand by on VHF Channel …/frequency … "

6.2.1 When it is accepted to remain on the VHF channel/ frequency indicated, say:

"Standing by on VHF Channel …/frequency … "

6.3 When it is advisable to change to another VHF Channel/ frequency, say:

"Advise (you) change to VHF Channel …/frequency … ."

"Advise(you) try VHF Channel …/frequency… ."

6.3.1 When the changing of a VHF Channel/frequency is accepted, say:

"Changing to VHF Channel …/frequency … ."

5.2 팡 팡 긴급 통보를 알리기 위해서 사용해야 한다.

5.3 시큐리티 안전 통보를 알리기 위해서 사용해야 한다.

6. 표준 준비 용어

6.1 "(이쪽) 신호는 잘 들립니까?"

6.1.1 **그쪽 신호의 감도는…**

매우 좋지 않습니다/1	신호의 강도는 1(즉, 겨우 들립니다)
좋지 않습니다/2	신호의 강도는 2(즉, 약합니다)
보통입니다/3	신호의 강도는 3(즉, 보통입니다)
좋습니다/4	신호의 강도는 4(즉, 좋습니다)
매우 좋습니다/5	신호의 강도는 5(즉, 매우 좋습니다)

6.2 어떤 한 VHF채널/주파수에서 대기하라고 권고하고자 할 경우에는 다음과 같이 말한다:

"VHF 채널…/주파수…에 대기하시오"

6.2.1 지시된 VHF 채널/주파수에 대기할 것을 수락하는 경우에는 다 음과 같이 말한다:

"VHF 채널…/주파수…에 대기하고 있겠음"

6.3 다른 VHF 채널/주파수로 변경하라고 권고하고자 할 경우에 는 다음과 같이 말한다:

"VHF채널…/주파수… (으)로 변경하시오"

"VHF채널…/주파수… (으)로 시도해보시오"

6.3.1 VHF채널/주파수 변경에 동의하고자 하는 경우에는 다음과 같 이 말한다:

"VHF채널…/주파수… 으로 변경하겠음"

7 Corrections

When a mistake is made in a message, say:

"Mistake …" - followed by the word:

"Correction … " plus the corrected part of the message.

Example: "My present speed is 14 knots - mistake 14. Correction, my present speed is 12, one-two, knots."

8 Readiness

"I am/I am not ready to receive your message".

9 Repetition

9.1 If any part of the message are considered sufficiently important to need safeguarding, say:

"Repeat …" - followed by the corresponding part of the message.

Example: "My draft is 12.6, repeat one-two decimal 6 metres."

"Do not overtake - repeat - do not overtake."

9.2 When a message is not properly heard, say:

"Say again (please)."

10 Numbers

Numbers are to be spoken in separate digits:

"One-five-zero" for 150

"Two decimal five" or

"Two point five" for 2.5

Note: Attention! When rudder angles e.g. in wheel orders are given, say:

"Fifteen" for 15 or "Twenty" for 20 etc..

7. 정 정

통보 사항에 오류가 발생된 경우에는 다음과 같이 말한다:

"오류…" 그 뒤에 그 단어를 말한다:

"정정…" 통보의 정정된 내용을 부가할 것.

보기: "본선의 현재 속력은 14노트 입니다-오류 14.

정정, 본선의 현재 속력은 12, one-two,노트"

8. 준 비

"이쪽은 그쪽의 통보를 받을 준비가 되어있음/있지 않음."

9. 반 복

9.1 통보 사항 중의 어떤 부분이 특히 중요하다고 여겨 보호할 필요가 있을 경우에는 다음과 같이 말한다:

"반복(repeat)…" 그리고 나서 통보 사항 중의 해당 부분을 말한다.

보기: "본선의 흘수는 12.6, 반복 12.6 미터입니다."

"추월하지 마시오 - 반복 - 추월하지 마시오."

9.2 통보 사항이 잘 들리지 않을 경우에는 다음과 같이 말한다:

"다시 말해 주시오"

10. 수

수는 한자씩 분리된 숫자로 말한다:

150은 "one-five-zero"로

2.5는 "Two decimal five" 또는 "Two point five" 로

주: 주의! 타각 즉 조타 명령을 말하고자 할 때는:

15는 "Fifteen"으로, 20은 "Twenty" 등으로 말한다.

11 Positions

11.1 When latitude and longitude are used, these shall be expressed in degrees and minutes(and decimals of a minute if necessary), North or South of the Equator and East or West of Greenwich.

Example: "WARNING. Dangerous wreck in position 15 degrees 34 minutes North 061 degrees 29 minutes West."

11.2 When the position is related to a mark, the mark shall be a well-defined charted object. The bearing shall be in the 360 degrees notation from true north and shall be that of the position FROM the mark.

Example: "Your position bearing 137 degrees from Big Head lighthouse distance 2.4 nautical miles."

12 Bearings

The bearing of the mark or vessel concerned, is the bearing in the 360 degree notation from north (true north unless otherwise stated), except in the case of relative bearings. Bearings may be either FROM the mark or FROM the vessel.

Examples: "Pilot boat is bearing 215 degrees from you."

Note: Vessels reporting their position should always quote their bearing FROM the mark, as described in paragraph 11.2.

12.1 Relative bearings

Relative bearings can be expressed in degrees relative to the vessel's head. More frequently this is in relation to the port or starboard bow.

Example: "Buoy 030 degrees on your port bow."

(Relative D/F bearings are more commonly expressed in the 360 degree notation.)

11. 위 치

11.1 위도와 경도를 사용하고자 할 때는 도 와 분(필요하면 분의 소수) 및 북위와 남위, 동경과 서경으로 나타낸다.

보기: "경고. 북위 15도 34분, 서경 061도 29분에 위험한 난파선 이 있음"

11.2 표지와 관련지어 위치를 나타내고자 할 경우, 그 표지는 해도 상에서 확실히 알 수 있는 것이어야 한다.

방위는 진북을 기점으로 하여 360도식으로 표시하고 표지로부터 나타낼 위치를 바라본 방위를 표시한다.

보기: "그쪽의 위치는 빅 헤드 등대로부터 137도, 거리 2.4해리 지점임."

12. 방 위

관련된 표지 또는 선박의 방위는 상대방위로 표시하는 경우를 제외하고는 북쪽(다른 단서가 없는 한 진북)을 기점으로 360도식으로 표시한다.

방위는 표지나 선박 중 어느 쪽으로부터의 방위로 표시해도 된다.

보기: "도선선은 귀선으로부터 215도 방향에 있음."

주: 본 장의 11.2 절에 표시된 바와 같이, 선박이 자선의 위치를 보고할 때에는 항상 표지(물표)에서 자선을 보는 방위로 나타내야 한다.

12.1 상대방위

상대방위는 선수를 기준으로 한 도수로 나타내고, 선수 좌현 또는 선수 우현 몇 도라고 표시하는 경우가 많다.

보기: "부표는 귀선의 선수 좌현 030도 방향에 있음."

(무선 방위 측정기에 의한 상대방위는 보통 360도식으로 표시한다.)

13 Courses

Always to be expressed in 360 degree notation from north (true north unless otherwise stated). Whether this is to TO or FROM a mark can be stated.

14 Distances

To be expressed in nautical miles or cables (tenths of a mile), the unit always to be stated.

15 Speed

To be expressed in knots:

15.1 without further notation meaning speed through the water; or,

15.1.1 "ground speed" meaning speed over the ground.

16 Times

Times should be expressed in the 24 hour UTC notation;[8] if local time will be used in ports or harbours it should clearly be stated.

17 Geographical names

Place names used should be those on the chart or in Sailing Directions in use. Should these not be understood, latitude and longitude should be given.

18 Ambiguous words

Some words in English have meanings depending on the context in which they appear. Misunderstandings frequently occur, especially in VTS communications, and have produced accidents. Such words are:

18.1 The Conditionals **"May", "Might", "Should"** and **"Could".**

<u>May</u>

Do not say: "May I enter the fairway?"

Say: "QUESTION. Do I have permission to enter the fairway?"

13 침 로

항상 북쪽(다른 단서가 없는 한 진북)을 기준으로 360도식으로 나타낸다. 이는 "어느 표지 방향으로" 또는 "어느 표지로부터" 의 표현 방법 중 어느 것인지를 언급해 줄 수 있다.

14 거 리

거리는 해리나 케이블(0.1해리)로 표시하며, 단위를 반드시 명시할 것.

15 속 력

노트로 표시한다:

15.1 특별한 단서가 없으면 대수속력을 나타내며; 또는,

15.1.1 "ground speed"라 하면 대지속력을 의미한다.

16 시 간

시간은 세계시 24시간 표시법으로 표시하고; 항구 내에서 지방시가 사용될 경우에는 그것을 명백하게 표시해야 한다.

17 지 명

사용된 지명은 사용 중인 해도나 수로지에 나와 있는 것이어야 한다.

이렇게 하여도 이해가 되지 않으면, 위도와 경도를 덧붙여야 한다.

18 애매한 어휘

영어에서 몇몇 어휘들은 그것들이 나타나는 문맥에 의하여 의미가 결정된다. 특히 VTS 통신 도중에 자주 오해가 발생되어 왔고, 이로 인하여 사고가 발생되어 왔다. 그러한 어휘는 다음과 같은 것들이 있다:

18.1 **"May", "Might", "Should" 및 "Could"**와 같은 가정 어귀.

May

다음과 같이 표현하지 말 것 : "May I enter the fairway?"

다음과 같이 표현할 것 : "QUESTION. Do I have permission to enter the fairway?"

Do not say: "You may enter the fairway."

Say: "ANSWER. You have permission to enter the fairway."

Might

Do not say: "I might enter the fairway."[9)]

Say: "INTENTION. I will enter the fairway."

Should

Do not say: "You should anchor in anchorage B3."

Say: "ADVICE. Anchor in anchorage B3."

Could

Do not say: "You could be running into danger."

Say: "WARNING. You are running into danger."

18.2 The word **"Can"**

The word "Can" either describes the possibility or the capability of doing something. In the SMCP the situations where phrases using the word "Can" appear make it clear whether a possibility is referred to. In an ambiguous context, however, say, for example:

"QUESTION. Do I have permission to use the shallow draft fairway at this time ? ",

do not say: "Can I use the shallow draft fairway at this time?",

if you ask for a permission. (The same applies to the word "May")

Note: In all cases the radiotelephone procedures as set out in the ITU - Radio Regulations have to be observed.[10)]

다음과 같이 표현하지 말 것: “You may enter the fairway.”

다음과 같이 표현할 것 : “ANSWER. You have permission to enter the fairway.”

Might

다음과 같이 표현하지 말 것 : “I might enter the fairway.”

다음과 같이 표현 할 것 : “INTENTION. I will enter the fairway.”

Should

다음과 같이 표현하지 말 것: “You should anchor in anchorage B3.”

다음과 같이 표현할 것 : “ADVICE. Anchor in anchorage B3.”

Could

다음과 같이 표현하지 말 것 : “You could be running into danger.”

다음과 같이 표현할 것 : “WARNING You are running into danger.”

18.2 **“Can”**이라는 단어

“Can”이라는 단어는 어떤 일의 실현 가능성이나 할 수 있는 능력을 나타낸다. 표준 해사 통신 용어에서는, “Can”이라는 단어를 쓰는 어귀가 나타나는 상황에서는 그것이 실현 가능성에 대하여 말하고 있는지를 명확하게 밝힌다.

그러나 애매한 문맥에서는, 허락을 요청할 경우 다음 예와 같이 표현하고,

“QUESTION. Do I have permission to use the shallow draft fairway at this time?”

다음과 같이 표현하지 않는다: “Can I use the shallow draft fairway at this time?” (“May”라는 단어에도 동일하게 적용됨)

주: 모든 경우에 있어서 ITU-무선 규칙에 상세히 설명된 무선전화 절차를 지켜야한다.

GLOSSARY

The GLOSSARY also includes a limited number of technical terms which do not appear in the text of the SMCP but might be useful in case the content of a given standard phrase requires modification.

1 General terms

Abandon vessel To evacuate crew and passengers from a vessel following a distress

Accommodation ladder Ladder attached to platform at vessel's side with flat steps and handrails enabling persons to embark/disembark from water or shore[11)]

Adrift Uncontrolled movement at sea under the influence of current, tide or wind

Air draft The height from the waterline to the highest point of the vessel

Assembly station Place on deck, in mess rooms, etc., assigned to crew and passengers where they have to meet according to the muster list when the corresponding alarm is released or announcement made

Backing (of wind) Shift of wind direction in an anticlockwise manner, for example from north to west(opposite of veering)

Beach (to) To run a vessel up on a beach to prevent its sinking in deep water

Berth
1: A sea room to be kept for safety around a vessel, rock, platform, etc.
2: The place assigned to a vessel when anchored or lying alongside a pier, etc.

용어 해설

용어 해설에는 SMCP 문장에 나타나지 않는 제한된 숫자의 기술 용어들이 포함되어 있는데, 이는 주어진 표준 어귀의 내용을 수정할 필요가 있는 경우에 유용할 것이다.

1 일반 용어

퇴선 조난으로 인하여 승무원과 여객을 선박으로부터 대피시킴

현문사다리 현측의 발판에 취부된 사다리로서 평평한 계단과 난간이 설치되어서 사람이 수면이나 육상으로부터 승/하선 할 수 있도록 해 주는 것

표류 해상에서 해류, 조류, 바람의 영향으로 제어되지 않고 떠다니는 것

높이 수면으로부터 선체의 가장 높은 부분까지의 높이

집합장소 구령이 발령되거나 해당 비상신호가 울렸을 때, 부서 배치표에 정하여진 바대로 승무원과 여객이 집합하도록 할당된 갑판상의 장소, 식당 등

(풍향)반전 풍향이 반시계방향, 예를 들면 북풍으로부터 서풍으로 변하는 것 (풍향 순전의 반대)

좌안 수심이 깊은 곳에 선박이 침몰하는 것을 방지하기 위하여 배를 해변에 끌어 올려놓음 (임의 좌초)

여유수면

1. 선박, 바위, 플랫폼의 주위에 안전을 위하여 (정박지) 확보해야 할 해상 여유 공간

2. 어느 선박의 투묘나 부두 계류 등을 위하여 할당된 장소

Blast A whistle signal made by the vessel

Blind sector An area which cannot be scanned by the ship's radar because it is shielded by parts of the superstructure, masts, etc.[12)]

Boarding arrangements All equipment, such as pilot ladder, accommodation ladder, hoist, etc., necessary for a safe transfer of the pilot

Boarding speed The speed of a vessel adjusted to that of a pilot boat at which the pilot can safely embark/ disembark

Bob-cat A mini-caterpillar with push-blade used for the careful distribution of loose goods in cargo holds of bulk carriers[13)]

Briefing Concise explanatory information to crew and/or passengers

Cable .1 Chain connecting a vessel to the anchor(s)

.2 Wire or rope primarily used for mooring a ship

.3 (Measurement), one hundred fathoms or one tenth of a nautical mile

Capsize To turnover

Cardinal buoy A seamark, i.e. a buoy, indicating the North, East, South or West, i.e. the cardinal points from a fixed point such as a wreck, shallow water, banks, etc.[14)]

Cardinal points The four main points of the compass: north, east, south and west

Casualty here: Case of death in an accident or shipping disaster

기적신호	선박의 기적 신호
맹목구간	선박의 상부 구조물, 마스트 등으로 인하여 차폐되어 선박의 레이더로 탐지할 수 없는 구역(차폐구역)
승선 설비	파일럿 사다리, 현문사다리, 호이스트 등 도선사의 안전한 승하선을 위하여 필요한 모든 장비
승선 속력	도선사가 안전하게 승/하선할 수 있도록 도선선의 속력에 맞춰 조정된 본선의 속력
보브캣	산적 화물선의 화물창에 있는 산적 화물을 주의 깊게 배치하기 위하여 사용되는 푸시블레이드가 장치된 소형 무한궤도 차
상황설명	승무원과 승객에 대한 간단한 설명적 통보
케이블	1. 닻과 선체를 연결하는 사슬
	2. 선박을 계류하는데 주로 사용되는 와이어나 밧줄
	3. (치수) 100 패덤 즉, 0.1 해리
전복	상하가 전도되는 것
방점부표	난파선, 여울, 사주 등, 정해진 지점의 북, 동, 남, 서, 즉 방점을 표시하는 항로표지 (방위표지)
방점	나침반의 네 개의 주요 방위: 즉, 북, 동, 남, 서
사고	여기서는: 사고나 해난으로 인한 사망사고를 말함

Check (to) 1: To make sure that equipment etc. is in proper condition or that everything is correct and safe

2: To regulate motion of a cable, rope or wire when it is running out too fast

Close-coupled towing A method of towing vessels through polar ice by means of icebreaking tugs with a special stern notch suited to receive and hold the bow of the vessel to be towed

Close up (to) To decrease the distance to the vessel ahead by increasing one`s own speed

Compatibility(of goods) states whether different goods can be safely stowed together in one cargo spaces[15)]

Vessel constrained by her draft A vessel severely restricted by her draught in her ability to deviate from the course followed in relation to the available depth and width of navigable water

Convoy A group of vessels which sail together, e.g. through a canal or ice

Course The intended direction of movement of a vessel through the water

Course made good That course which a vessel makes good over ground, after allowing for the effect of cur-rents, tidal streams, and leeway caused by wind and sea[16)]

COW Crude Oil Washing: A system of cleaning the cargo tanks by washing them with the cargo of crude oil during discharge[17)]

CPA/TCPA Closest Point of Approach/Time to Closest Point

점검	1. 장비 등이 적절한 상태에 있는지, 또는 모든 것이 제대(억제)로 되어 있고, 안전한지를 확인하는 것 2. 닻줄, 밧줄, 와이어 등이 너무 빨리 풀려 나갈 때 그 움직임을 억제하는 것
밀접결합예인	피예인선의 선수를 받아 들여 잡아 맬 수 있도록 선미에 특별한 홈이 설치된 쇄빙 예인선을 사용하여 수행하는 극지방 빙해역에서의 선박 예인 방법
접근	자선의 속력을 증가시켜 전방의 선박과의 거리를 좁히는 것
격리 특성	서로 다른 화물을 동일한 화물 공간에 같이 안전하게 적(화물의)부할 수 있는지의 여부
흘수 제한선	항행 가능한 해수의 수심과 폭에 관련하여, 선박의 흘수 때문에 그 선박이 따라 가야만 할 침로로부터 벗어날 수 있는 능력이 현저하게 제한되는 선박
호송 선단	운하나 빙해를 함께 항해하는 한 떼의 선박들
침로	수면을 기준으로 한, 어느 선박이 이동하고자 하는 방향
대지 침로	해류, 조류, 바람과 파도에 의한 풍압편위 등의 영향을 고려하여 대지를 기준으로 한 침로
COW	원유 양하 후에 화물인 원유로 화물탱크를 세척하는 시스템 (원유 세정)
CPA/TCPA	최접근점/추적하고 있던 표적이 그러한 한계 이내로

of Approach limit as defined by the obser-ver to give warning when a tracked target or targets will close to within these limits

Crash-stop An emergency reversal operation of the main engine(s) to avoid a collision

Damage control team A group of crew members trained for fighting flooding in the vessel

Datum
1. The most probable position of a search target at a given time
2. The plane of reference to which all data as to the depth on charts are referenced.[18)]

Derelict Vessel still afloat, abandoned at sea

Destination Port which a vessel is bound for

Disabled A vessel damaged or impaired in such a manner as to be incapable of proceeding on its voyage

Disembark (to) To go from a vessel

Distress alert(GMDSS) A radio signal from a distressed vessel automatically directed to an MRCC giving position, identification,[19)] course and speed of the vessel as well as the nature of distress

Distress/Urgency traffic here: The verbal exchange of information on radio from ship to shore and/or ship to ship/air craft about a distress/urgency situations defined in the relevant ITU Radio Regulations

Draught (or draft) Depth in water at which a vessel floats

Dragging (of anchor) Moving of an anchor over the sea bottom involuntarily because it is no longer preventing the movement of the vessel

	접근하게 되면 경보를 발하도록 관측자가 설정한 최접근 한계에 표적 도달하는 시간
긴급 정지	충돌을 피하기 위하여 주기관을 비상 역전시킴
손상제어반	선내의 침수 방지에 대한 훈련을 받은 승무원 집단
추정 기점	1. 주어진 시각에 있어서 목표물을 발견할 가능성이 가장 높은 위치 2. 해도상의 수심에 관한 모든 자료의 기준이 되는 기준면. (기본 수준면)(약최저저조면)
유기선	해상에 버려진, 떠있는 선박
목적지	어떤 선박이 향하는 목적지 항구
파손 선박	항해를 계속할 수 없을 정도로 파손되거나 장애가 있는 선박
하선	선박으로부터 내려옴
조난경보 (GMDSS)	조난선으로부터 자동적으로 해상 구조 조정 본부(MRCC)에 보내지는 위치, 식별부호, 침로, 속력 및 조난의 성질 등을 포함하는 무선 신호
조난/긴급 통화	여기서는: ITU의 관련 무선 규칙에 정의된 조난/긴급 상황에 대한 선박과 육상간/선박과 선박간/선박과 항공기간의 무선 음성 정보 교환
흘수	물에 뜬 선체의 깊이
주묘(닻의)	닻이 선체의 이동을 더 이상 저지하지 못하여 어쩔 수 없이 해저를 따라 끌리고 있는 상태

Dredging (of anchor) Moving of an anchor over the sea bottom to control the movement of the vessel[20)]

Drifting Being driven along by the wind, tide or current

Drop back (to) To increase the distance to the vessel ahead by reducing one's own speed

DSC Digital Selective Calling (in the GMDSS system)[21)]

Embark (to) To go aboard a vessel

EPIRB Emergency Position Indicating Radio Beacon[22)]

Escape route A clearly marked way in the vessel which has to be followed in case of an emergency

Escort Attending a vessel, to be available in case of need, e.g. ice-breaker, tug, etc.

ETA Estimated Time of Arrival

ETD Estimated Time of Departure

Fathom A measure of 6 feet

Fire patrol A member of the watch going around the vessel at certain intervals so that an outbreak of fire may be promptly detected; mandatory in vessels carrying more than 36 passengers

Flooding Major uncontrolled flow of seawater into the vessel

Fire monitor Fixed foam/powder/water cannon shooting fire extinguishing agents on tank deck, manifold etc.

Foul (of anchor) Anchor has its own cable twisted around it or has fouled an obstruction

드레징(닻의) 선체의 운동을 제어하기 위하여 해저를 따라서 닻을 이동 시키는 상태

표류 바람, 조류, 해류에 의하여 떠내려감

뒤처짐 자선의 속력을 저하시킴으로써 앞에 있는 선박과의 거리를 넓힘

DSC 디지털 선택 호출(GMDSS 시스템 내의)

승선 선박에 탑승함

EPIRB 비상 위치 표시용 무선 비이콘

탈출 통로 선박에서 비상사태가 발생할 때 따라가야 할 대피 통로로서 명확하게 표시된 길

호송 필요한 경우 이용할 수 있도록 어떤 선박에 동행함 즉, 쇄빙선, 예인선 등

ETA 도착 예정 시각

ETD 출발 예정 시각

패덤 6 피트의 길이

화재순찰자 화재의 발생을 조기에 탐지할 수 있도록 일정 간격으로 선내를 순찰하는 당직원; 36인 이상의 여객을 수송하는 선박에서는 강제 사항임

침수 해수가 제어되지 않고 대량으로 선내에 유입됨

소화 모니터 탱크의 갑판이나 매니폴드 등에 소화 약제를 발사하는 고정식의 포말/분말/소화수 발사기

엉킴(닻) 닻이 그 닻줄에 감기거나 장애물에 걸림

Foul (of propeller) A line, wire, net, etc., is wound round the propeller

Full speed Highest possible speed of a vessel

Fumes Often harmful gas produced by fires, chemicals, fuel, etc.

General emergency alarm A sound signal of seven short blasts and one prolonged blast given with the vessel's sound system

Give way To keep out of the way of another vessel

GMDSS Global Maritime Distress and Safety System[23)]

(D) GPS (Differential) Global (satellite) Positioning System[24)]

Half cardinal points The four main points lying between the cardinal points: north east, south east, south west and north west

Hampered vessel A vessel restricted by her ability to manoeuvre by the nature of her work[25)]

Hatchrails Ropes supported by stanchions around an open hatch to prevent persons from falling into a hold

Heading The horizontal direction of the vessel's bows at a given moment measured in degrees clockwise from north

Hoist here: A cable used by helicopters for lifting or lowering persons in a pick-up operation

Icing Coating of ice on an object, e.g. the mast or superstructure of a vessel

IMO-Class Group of dangerous or hazardous goods, harmful substances or marine pollutants in sea transport as classified in the International Maritime Dangerous Goods Code (IMDG Code)[26)]

엉킴(프로펠러) 밧줄, 와이어, 그물 등이 프로펠러에 엉킴

전속력 어떤 선박의 가장 빠른 속력

유해가스 화재, 케미컬, 연료 등에 의해서 생기는 해로운 가스

일반비상경보 선박의 음향신호장치로 발생시키는 단성 7발, 장성 1발로 이루어지는 음향신호

피항 다른 선박의 진로로부터 비켜남 (양보)

GMDSS 전 세계 해상 조난 및 안전 체계

(D)GPS (차분-보정) 전세계 (위성)위치 결정 시스템

우점 방점 사이에 있는 네 주요점: 북동, 남동, 남서, 북서

조종성능제한선 종사하고 있는 작업의 성격상 조종성능이 제한 받고 있는 선박

해치레일 사람이 화물창 안으로 추락하는 것을 방지하기 위하여 개방된 해치 주위에 지주로 지지하여 설치되는 밧줄

선수 방위 어느 주어진 순간에 북쪽으로부터 선수의 방향까지 시계방향으로 측정한 수평각

호이스트 여기서는: 헬리콥터에서 인양 작업중 사람을 들어 올리거나 내릴 때 사용하는 밧줄 (권양기)

착빙 선박의 마스트, 상부구조물과 같은 물체에 얼음이 얼어서 덮는 것

IMO 등급 해상으로 운송되는 위험화물, 유해물질, 해양오염 물질의 부류로서 IMDG Code 로 분류된 것

Inert (to) To reduce the oxygen in a tank by inert gas to avoid an explosive atmosphere[27)]

Initial course Course directed by the OSC or other authorized person to be steered at the beginning of a search

Inoperative Not functioning

Jettison(to)(of cargo) Throwing overboard of goods in order to lighten the vessel or improve its stability in case of an emergency

Launch (to) To lower, e.g. lifeboats to the water

Leaking Escape of liquids such as water, oil, etc., out of pipes, boilers, tanks, etc., or a minor inflow of seawater into the vessel due to damage to the hull

Leeward On or towards the sheltered side of a ship; opposite of windward

Leeway Vessels sideways drift leeward of the desired course[28)]

Let go (to) To set free, let loose, or cast off (of anchors, lines, etc.)

Lifeboat station Place assigned to crew and passengers to muster before being ordered into the lifeboats

List here: Inclination of the vessel to port side or starboard side

Located In navigational warnings: Position of object confirmed

불활성(화) 불활성 기체를 탱크에 주입하여 산소를 제거하므로서, 폭발을 방지하는 것

최초 침로 현장 지휘자나 다른 권한 있는 자가 지시한, 수색 개시 당시의 조타 침로

작동되지 않는 기능을 발휘하고 있지 않는

투하(화물의) 비상시에 선박을 가볍게 하거나 복원성을 향상시키기 위하여 화물을 선외로 버리는 것

진수 강하 즉, 구명정을 수면에 하강시킴

누출 물, 유류 등의 액체가 파이프, 보일러, 탱크 등으로 부터 누출되는 것. 또는 선체의 손상으로 인하여 소량의 해수가 선내로 유입되는 것

풍하측 선체에 의하여 바람으로부터 차폐되는 방향 또는 그 쪽 현; 풍상 측의 반대

풍압 편위 선박이 그 의도하는 침로로부터 이탈하여 바람이 불어가는 쪽으로 밀리는 것

Let go(to) 자유롭게 하다, 느슨하게 하다, 풀다(닻, 계류삭 등)

구명정 부서 구명정에 탑승하도록 지시 받기 위하여 승무원과 여객이 집합할 장소로 할당된 곳

횡경사 여기서는 : 좌현쪽이나 우현쪽으로 선박이 경사함

위치 확인 항행 경보에서 : 목표물의 위치를 확인함

Make water (to) Seawater flowing into the vessel due to hull damage, or hatches awash and not properly closed

MMSI Maritime Mobile Service Identity number[29)]

Moor (to) To secure a vessel in a particular place by means of wires or ropes made fast to the shore, to anchors, or to anchored mooring buoys, or to ride with both anchors down

MRCC Maritime Rescue Co-ordination Centre: Land-based authority responsible for promoting efficient organization of maritime search and rescue and for co-ordinating the conduct of search and rescue operations within a search and rescue region[30)]

Muster (to) To assemble crew, passengers or both in a special place for purposes of checking

Muster list List of crew, passengers and others on board and their functions in a distress or drill

Not under command (abbr. NUC) A vessel which through exceptional circumstances is unable to manoeuvre as required by the COLREGs[31)]

Obstruction An object such as a wreck, net, etc., which blocks a fairway, route, etc.

Off air When the transmissions of a radio station etc., have broken down, been switched off or suspended

Off station (of buoys) Not in charted position

Oil clearance Oil skimming from the surface of the water

Operational Ready for immediate use

누수 선체의 손상으로 인하거나, 해치가 적절히 밀폐되지 못하고 물에 잠겨서 해수가 선내로 침입함

MMSI 해상 이동 전화 식별 번호 (p. 173)

계류 선박을 육상, 닻, 계류용 부표 등에 고정된 와이어나 밧줄에 의하여 어떤 특정한 장소에 고정시킴, 또는 쌍묘박을 일컬음

MRCC 해상 구조 조정 본부: 해상 수색 구조 작업의 효과적인 조직을 도모하고, 수색 구조 구역 내에서 수색구조 작업의 지휘를 조정할 책임이 있는 육상의 당국지

소집 점검을 위하여 승무원, 여객 또는 모두를 특정의 장소에 집합시킴

부서 배치표 조난시, 훈련시에 대한 선내의 승무원, 여객, 기타 인원의 명단과 그들의 직무 일람표

운전부자유선 예외적인 상황으로 인하여 COLREG에서 요구하는 바대로 조종할 수 없는 선박 (약어 : NUC)

장애물 난파선, 그물 등과 같이 항로를 가로막는 물체

방송 중지 무선 방송국 등의 발신이 고장나거나, 꺼졌거나, 일시 중지된 상태

위치이탈(부표) 해도에 기재된 위치에 있지 않음

유류 제거 수면에 있는 유류를 제거하는 작업

작동 가능 즉시 사용할 수 있음

Ordnance exercise Naval firing practice

OSC On-Scene Co-ordinator: A person designed to coordinate search and rescue operations within a specified area

Overflow Escape of oil or liquid from a tank because of a two-fold condition as a result of overflowing, thermal expansion, change in vessel trim or vessel movement

Polluter A vessel emitting harmful substances into the air or spilling oil into the sea

Preventers Ropes or wires attached to derricks to prevent them from swinging during cargo handling operations

Proceed (to) To sail or head for a certain position or to continue with the voyage

PA-system Public address system: Loudspeakers in the vessel's cabins, mess rooms, etc., and on deck through which important information can be broadcast from a central point, mostly from the navigation bridge

Recover (to) Here: To pick up shipwrecked persons

Refloat (to) To pull a vessel off after grounding; to set afloat again

Rendezvous An appointment between vessels normally made on radio to meet in a certain area or position

Reported in navigational warnings: Position of object unconfirmed

Restricted area A deck, space, area, etc., in vessels, where for safety reasons, entry is only permitted for authorized crew members

포격 훈련 해군의 사격 훈련

OSC 현장 지휘자: 특정 해역에서 수색 구조 작업을 조정 하도록 지정된 자

넘침 넘침, 열팽창, 선박의 트림 변화, 선체 운동에 따른 이중적인 조건에 의하여 유류나 액체가 탱크로부터 빠져 나감

오염원 선박 유해 물질을 대기중에 방출하거나 해상에 유류를 유출하는 선박

프리벤터 하역 작업 중에 데리크가 선회하지 않도록 데리크를 묶어둔 밧줄이나 와이어

항진 특정한 지점으로 향하거나 항해함, 항해를 계속함

PA장치 확성방송장치: 대개 항해 선교 등의 중앙 위치로부터 중요한 통보 사항을 방송할 수 있도록 선박의 선실, 식당 등과 갑판에 설치된 확성기

인양 여기서는: 난파된 사람을 끌어올림

재부양 좌초 후 선박을 예인하여 이초시킴: 재부양시킴

랑데부 어떤 특정한 해역이나 위치에서 상호 만나기로, 선박 사이에 보통 무선으로 실시하는 약속

보고됨 항행 경보에서: 물체의 위치가 확인되지는 않았음

제한구역 선내에서 안전상의 이유로 권한이 부여된 승무원에게만 진입이 허락된 갑판, 장소, 구역 등

Resume (to) here: To re-start a voyage, service or search

Retreat signal Sound, visual or other signal to a team ordering it to return to its base

Rig move The movement of an oil rig, drilling platform, etc., from one position to another

Roll call The act of checking who of the passengers and crew members are present, e.g. at assembly stations, by reading aloud a list of their names

Safe speed That speed of a vessel allowing time for effective action to be taken under prevailing circumstances and conditions to avoid a collision and to be stopped within an appropriate distance[32)]

SWL Safe working load: maximum working load of lifting equipment that should not be exceeded

Safe working pressure The maximum permissible pressure in cargo hoses

SAR Search and Rescue

SART Search and Rescue Transponder

Scene The area or location where the event, e.g. an accident has happened

Search pattern A pattern according to which vessels and/or aircraft may conduct a co-ordinated search (the IMOSAR offers seven search patterns)[33)]

Search speed The speed of searching vessels directed by the OSC

Seamark A navaid placed to act as a beacon, or warning

재개	여기서는: 항해, 업무, 또는 수색을 다시 시작함
철수신호	그 작업조의 원래 기지로 돌아가라는 음향, 시각 기타의 신호
시추선 이동	유류 시추선, 시추 플랫폼 등이 한 장소로부터 다른 장소로 이동함
인원점검	집합 장소에서 명부를 크게 호명함으로써 승무원이 나 승객의 출석 상태를 점검함
안전속력	주어진 상황과 조건에서 충돌을 피하기 위한 효과적인 행동을 취할 수 있는 시간을 확보하고 적절한 거리 이내에서 선박이 정지할 수 있는 속력
SWL	안전 작업 하중: 인양 설비의 최대 작업 하중으로서 이를 초과해서는 안됨
안전작업압력	화물용 호스의 최대 허용 압력
SAR	수색과 구조
SART	수색구조용 트랜스폰더
현장	사고가 발생한 지역이나 위치
수색방식	선박 및/또는 항공기 상호간의 협조수색시 채택할 수색방식 (IMOSAR에는 일곱 가지의 수색 방식이 제시되어 있다)
수색 속력	현장 지휘자 (OSC)에 의하여 지시된 수색선의 속력
해상 표지	비이컨이나 경고표지의 역할을 하도록 설치된 항행 보조장치

Segregation(of goods) Separation of goods which for different reasons must not be stowed together

Shackle .1 Length of chain cable measuring 15 fathoms,
.2 U-shaped link closed with a pin used for connecting purposes

Shifting cargo Transverse movement of cargo, especially bulk, caused by rolling or a heavy list

Slings Ropes, nets, and any other means for handling general cargoes

Speed of advance The speed at which a storm centre moves

Spill (to) The accidental escape of oil, etc., from a vessel, container, etc., into the sea

Spill control gear Anti-pollution equipment for combating accidental spills of oils or chemicals[34)]

Elongated spreader here: Step of a pilot ladder which prevents the ladder from twisting[35)]

Stand by (to) To be in readiness or prepared to execute an order; to be readily available

Stand clear (to) here: To keep a boat away from the vessel

Standing orders Orders of the Master to the officer of the watch which s/he must comply with

Stand on (to) To maintain course and speed

Station The allotted place or the duties of each person on board

Stripping Final pumping of tank's residues[36)]

Survivor A person who continues to live in spite of being in an extremely dangerous situation, e.g. a shipping disaster.

격리(화물의)	서로 다른 이유로 인하여 같이 적재해서는 안되는 화물의 격리
샤클	1. 15 패덤의 닻줄 길이 2. 핀으로 잠기게 되어 연결용으로 쓰이는 U자 모양으로 생긴 고리
화물 이동	특히 산적 화물에 있어서, 선박의 횡요나 큰 횡경사로 인하여 화물이 횡방향으로 이동하는 것
슬링	잡화를 취급할 때 사용하는 밧줄, 그물 등의 도구
전진 속도	폭풍 중심의 이동 속도
누출	선박, 용기 등으로부터 유류 등이 사고로 해상에 빠져나감
누출제어기구	사고로 누출된 유류/화학 약품을 제거하는 방제기구
연장 스프레이더	여기서는: 도선사용 사다리에서 사다리의 꼬임을 방지하도록 설치된 발판
Stand by(to)	명령을 실행할 대기 상태에 있음; 즉시 이용가능함
벗어나기	여기서는: 보트를 선박으로부터 떨어진 곳에 둠
당직근무수칙	선장이 당직 항해사에게 내리는 지시로서 당직사관이 지켜야 할 사항
유지	선박의 침로와 속력을 유지함
부서	선내의 각 인원에게 할당된 장소 혹은 임무
스트리핑	탱크의 잔유를 최종적으로 퍼냄
생존자	극히 위험한 상황 즉, 해난사고 중에서도 생명을 유지하고 있는 사람

Take off (to) A helicopter lifts off from a vessel's deck

Target The echo generated e.g. by a vessel on a radar screen

Tension winch A winch which applies tension to mooring lines to keep them tight

TEU Twenty Foot Equivalent Unit (standard container dimension)

Track The path followed, or to be followed, between one position and another

Transit here: The passage of a vessel through a canal, fairway, etc.

Transit speed Speed of a vessel required for the passage through a canal, fairway, etc.

Transhipment(of cargo) here: The transfer of goods from one vessel to another outside harbours

Underway A vessel which is not at anchor, or made fast to the shore, or aground[37)]

Union purchase A method of cargo handling by combining two derricks, one of which is fixed over the hatch, the other over the ship's side[38)]

Unlit When the light of a buoy or a lighthouse are inoperative

UTC Universal Time Co-ordinated (GMT)

Variable (of winds) A wind that is constantly changing speed and direction

Veering (of winds) Clockwise change in the direction of the wind; opposite of backing

Veer out(to)(of anchors) To let out a greater length of cable

이륙 선체의 갑판으로부터 헬리콥터가 이륙함

표적 레이더 화면상에 나타난 타선의 영상

장력 윈치 계류삭이 팽팽해지도록 계류삭에 장력을 주는 윈치

TEU Twenty Foot Equivalent Unit(표준 컨테이너 규격)

Track(진로) 한 위치로부터 다른 위치까지 이동한/ 이동할 경로

통과 여기서는: 선박의 운하, 항로 등의 통과

통과 속력 선박이 운하, 항로 등을 통과할 때 지켜야 할 속력

이적(화물) 여기서는: 항구 밖에서, 화물을 한 선박으로부터 다른 선박으로 옮겨 싣는 것

항해중 선박이 투묘, 계류, 좌초 상태에 있지 않은 것

Union Purchase 두 개의 데리크를 조합하여 행하는 하역방법으로서 하나의 데리크는 해치의 상방에 고정시키고 다른 하나는 현측의 상방에 고정시킴

소등 등대나 부표의 등화가 점등되어 있지 않은 것

UTC 세계 표준시(GMT)

변풍(바람) 풍속, 풍향이 끊임없이 변화하는 바람

(풍향)순전 풍향이 시계방향으로 변함; 풍향반전의 반대

VHF Very High Frequency (30 - 300 MHz)

Walk out (to) (of anchors) To reverse the action of a windlass to lower the anchor until it is clear of the hawse pipe and ready for dropping

Walk back (to) (of anchors) **To reverse the action of a windlass to ease** the cable

Way point A position a vessel has to pass or at which she has to alter course according to her voyage plan[39)]

Windward The general direction from which the wind blows; opposite of leeward

Wreck A vessel which has been destroyed or sunk or abandoned at sea

2 VTS special terms

Fairway Navigable part of a waterway

Fairway speed **Mandatory speed in a fairway**

ITZ Inshore Traffic Zone (of a TSS): A routing measure comprising a designated area between the landward boundary of a TSS and the adjacent coast[40)]

Manoeuvring speed A vessel's reduced speed in circumstances where it may be required to use the engines at short notice[41)]

Receiving point A mark or place at which a vessel comes under obligatory entry, transit, or escort procedure

Reference line A line displayed on the radar screens in VTS Centres and/or electronic sea-charts separating the fairway for inbound and outbound vessels so that they can safely pass each other

풀어주기(닻) 닻줄을 더 길게 풀어 주는 것

VHF 초단파(30-300 MHz)

Walk out(to) (닻의) 양묘기를 역전시켜 닻을 풀어내려 호스파이프로부터 벗어나게 하여 투묘 준비 상태로 하는 것

Walk back(to) (닻의) 닻줄을 늦춰주기 위하여 양묘기를 역전시킴

통과지점 어떤 선박이 통과해야 될 지점 또는 어떤 선박의 항해계획에 따라 변침해야 될 지점

풍상측 바람이 불어오는 방향; 풍하측의 반대

난파선 파괴되었거나, 침몰되었거나, 바다에 버려진 선박

2 VTS 특수 용어

항로 수로의 항행 가능한 부분

항로내 제한속력 항로 내에서 강제적으로 지켜야할 속력

ITZ 연안통항대(통항분리방안의) : 통항분리방안의 육지 쪽 경계와 인접한 해안선 사이에 있는 지정된 해역 을 포함하는 항로 지정 방안

조종속력 기관을 즉시 사용하게 될 수도 있는 상황 하에서의 감소된 선박 속력

지시 수령점 선박의 진입, 통과, 호송 작업에 있어서 강제적인 절차를 따르기 시작하는 지점 또는 그 지점을 나타내는 표지

참조선 VTS 센터의 레이더 화면이나 전자해도상에 표시되어 입항선과 출항선의 항로를 분리함으로써, 그들이 상호간에 안전하게 항과 할 수 있도록 해주는 선

Reporting point A mark or position at which a vessel is required to report to the local VTS-Station to establish its position

Separation zone/line
A zone or line separating the traffic lanes in which vessels are proceeding in opposite or nearly opposite directions; or separating a traffic lane from the adjacent sea area; or separating traffic lanes designated for particular classes of vessels proceeding in the same direction[42)]

Traffic clearance
VTS authorization for a vessel to proceed under conditions specified

Traffic lane An area within defined limits in which one-way traffic is established

TSS Traffic Separation Scheme: A routing measure aimed at the separation of opposing streams of traffic by appropriate means and by the establishment of traffic lanes[43)]

VTS Vessel Traffic Services: Services, designed to improve safety and efficiency of vessel traffic and to protect the environment[44)]

VTS-area Area controlled by a VTS-Centre or VTS-Station

보고지점 어떤 선박이 위치를 확인해 주기 위하여 지역의 VTS 국에 보고해야 하는 항해표지나 지점

분리대/선 서로 정반대 방향 또는 거의 반대 방향으로 선박이 진행하는 각각의 일방통항로를 서로 분리하거나; 일방통항로와 인접한 해역을 분리하거나; 같은 방향으로 선박이 진행하지만 각각 특정한 등급의 선박에 할당된 일방통항로들을 서로 분리하는 구역이나 선

통항 허가 선박으로 하여금 명시된 조건하에서 항진하도록 VTS가 허가하는 것

일방통항로 일방통항이 설정된 명시된 경계내의 해역

TSS 통항 분리 방안 : 일방통항로의 설정과 적절한 수단에 의하여 서로 반대 방향으로 진행하는 통항선의 흐름을 분리하고자 하는 항로 지정 방안

VTS 선박 통항 관제 업무: 선박 통항의 안전성, 효율성을 향상시키고 환경을 보호하기 위하여 설정된 업무

VTS 구역 VTS 본부나 VTS국의 통제를 받는 구역

STANDARD MARINE COMMUNICATION PHRASES: PART A

PART A covers Phrases applicable in external communications from ship to shore, shore to ship and ship to ship as required by STCW 1978, as revised, Table A-II/1, as well as Phrases applicable on board vessels in conversations between Pilots and bridge teams as required by Regulation 14(4) of Chapter V of SOLAS 1974, as revised.

AI EXTERNAL COMMUNICATION PHRASES

Attention: The use of Standard Phrases in vessels' external communi-cation does not in any way exempt from applying the radiotelephone procedures as set out in the ITU - Radio Regulations.

AI/1 Distress Traffic

The distress traffic controlling station/other stations may impose radio silence on any interfering stations by using the term:

"Seelonce Mayday/Distress"
unless the latter have messages about the distress.[45)]

AI/1.1 Distress Communications

Note: A distress traffic has always to commence with stating the position of the vessel in distress as specified in "GENERAL 11 Positions/12 Bearings" if it is not included in the DSC distress alert.

.1 Fire, explosion

.1 I am/MV ··· on fire (- after explosion).

.2 Where is the fire?

.2.1 Fire is
~ on deck.
~ in engine-room.

표준 해사 통신 용어: A편

A편에는 개정된 1978년 STCW의 표A-II/1에 의하여 요구되는 것으로서, 선박에서 육상으로, 육상에서 선박으로, 선박에서 선박으로의 외부 통신에 적용되는 용어와, 개정된 1974년 SOLAS 제5장 14(4) 규칙에 의하여 요구되는 것으로서, 도선사와 선교팀 사이의 회화에 적용되는 용어를 포함한다.

AI　외부 통신 용어

주의: 선박의 외부 통신에 있어서 표준 용어를 사용한다고 해도, ITU 전파 법규에 있는 무선 전화에 관한 절차의 적용을 면제하는 것은 전혀 아니다.

AI/1　조난 통화

조난 통화 제어국/기타국은:

"Seelonce Mayday/Distress" unless the latter have messages about the distress. 라는 용어를 사용하여 방해국에게 전파침묵을 요구할 수 있다.

AI/1.1　조난 통신

주: 조난선의 위치가 DSC 조난 경보에 포함되지 않았으면, "일반 사항 11 위치/12 방위"에 명시된 바대로 항상 조난선의 위치를 말하면서 조난 통화를 시작해야 한다.

.1　화재, 폭발

.1　본선에/기선 … 에 (폭발 후)화재가 발생함.

.2　어느 부분에 화재가 발생했는가?

.2.1　갑판에 화재 발생함.

기관실에 화재 발생함.

~ in hold(s).
~ in superstructure/accommodation/… .

.3 Are dangerous goods on fire?
.3.1 Yes, dangerous goods are on fire.
.3.2 No, dangerous goods are not on fire.
.4 Is there danger of explosion?
.4.1 Yes, danger of explosion.
.4.2 No danger of explosion.
.5 I am/MV … not under command.
.6 Is the fire under control?
.6.1 Yes, fire is under control.
.6.2 No, fire is not under control.
.7 What kind of assistance is required?
.7.1 I do not/MV … does not require assistance.
.7.2 I require/MV … requires

~ fire fighting assistance.
~ breathing apparatus - smoke is toxic.[46)]
~ foam extinguishers/CO_2 extinguishers.[47)]
~ fire pumps.
~ medical assistance/… .

.8 Report injured persons.
.8.1 No persons injured.
.8.2 Number of injured persons/casualties: … .

.2 Flooding

.1 I am/MV … is flooding below water line.
.2 I/MV … cannot control flooding.[48)]
.3 What kind of assistance is required?
.3.1 I require/MV … requires pumps/divers/… .
.3.2 I will send pumps/divers/… .
.3.3 I cannot send pumps/divers/… .

화물창에 화재 발생함.
상부 구조물/거주구역/… 에 화재 발생함.

.3 위험 화물에 화재가 발생했는가?
.3.1 예, 위험화물에 화재 발생함.
.3.2 아니오, 위험화물에 화재 발생하지 않았음.
.4 폭발의 위험성이 있는가?
.4.1 예, 폭발의 위험성이 있음.
.4.2 폭발의 위험성이 없음.
.5 본선/MV … 은 운전이 부자유함.
.6 불길은 잡히고(진압되고) 있는가?
.6.1 예, 불길은 잡히고(진압되고) 있음.
.6.2 아니오, 불길은 잡히지 않고 있음.
.7 어떤 종류의 원조를 필요로 하는가?
.7.1 본선/MV…은 원조를 필요로 하지 않음.
.7.2 본선/MV…은 소화 작업 원조를 필요로 함.
본선/MV은 자장식 호흡구를 필요로 함 - 연기가 유독성임.
본선/MV…은 포말 소화기/탄산가스 소화기를 필요로 함.
본선/MV…은 소화 펌프를 필요로 함.
본선/MV…은 의료 원조/…를 필요로 함.
.8 부상자에 대하여 보고하시오.
.8.1 부상자 없음.
.8.2 부상자/사상자 수는: …명임.

.2 침수

.1 본선/MV…은 수선 하부에서 물이 새고 있음.
.2 본선/MV…은 침수를 억제할 수 없음.
.3 어떤 종류의 원조를 필요로 하는가?
.3.1 본선/MV…은 펌프/잠수부/…가 필요함.
.3.2 이쪽은 펌프/잠수부/…를 보내겠음.
.3.3 이쪽은 펌프/잠수부/…를 보낼 수 없음.

.4 I have/MV … has dangerous list to port/starboard.
.5 I am/MV … in critical condition.
.6 Flooding is under control.
.7 I/MV … can proceed without assistance.
.8 I require/MV … requires escort/tug assistance/… .

.3 Collision

.1 I have/MV … has collided
~ with MV… .
~ with unknown vessel/object/… .
~ with …(*name*) light vessel.
~ with seamark …(*charted name*).
~ with iceberg/ … .
.2 Report damage.
.2.1 I have/MV … has damage above/below water line.
.2.2 I am/MV … not under command.
.2.3 I/MV … cannot establish damage.
.2.4 I/MV … cannot repair damage.
.2.5 I/MV … can only proceed at slow speed.
.3 What kind of assistance is required?
.3.1 I require/MV … requires escort/tug assistance/… .

.4 Grounding

.1 I am/MV … aground.
.2 I require/MV … requires tug assistance/pumps/… .
.3 What part of your vessel is aground?
.3.1 Aground forward/amidships/aft/full length.
.3.2 I cannot establish which part is aground.
.4 Warning. Uncharted rocks in position … .
.5 Risk of grounding at low water.
.6 I/MV … will jettison cargo to refloat.
.6.1 Warning! Do not jettison IMO-Class … cargo!

.4 본선/MV…은 위험하게 좌현/우현으로 횡경사 됨.
.5 본선/MV…은 중대한 상황에 처해 있음.
.6 침수는 진압되었음.
.7 본선/MV…은 원조 없이 항진할 수 있음.
.8 본선/MV…은 호송/예인선의 원조/…을 필요로 함.

.3 충 돌

.1 본선/MV…은 기선…와 충돌하였음.
본선/MV…은 미상의 선박/물체/…와 충돌하였음.
본선/MV…은 등선…(선명) 와 충돌하였음.
본선/MV…은 해상 표지 …(해도상의 명칭)와 충돌하였음.
본선/MV…은 빙산/…와 충돌하였음.
.2 피해 상황을 보고하시오.
.2.1 본선/MV…은 수선 상부/하부에 손상을 입었음.
.2.2 본선/MV…은 운전이 부자유함.
.2.3 본선/MV…는 손상을 파악할 수 없음.
.2.4 본선/MV…은 손상을 수리할 수 없음.
.2.5 본선/MV…은 오로지 저속으로만 항진할 수 있음.
.3 어떤 종류의 원조를 필요로 하는가?
.3.1 본선/MV…은 호송/예인선의 원조/…을 필요로 함.

.4 좌 초

.1 본선/MV…은 좌초되었음.
.2 본선/MV…은 예인선의 원조/펌프/…를 필요로 함.
.3 귀선의 어느 부분이 좌초되었는가?
.3.1 선수 부분/선체 중앙부/선미 부분/전 선체가 좌초되었음.
.3.2 어느 부분이 좌초되었는지 파악할 수 없음.
.4 경고. 해도에 기재되어 있지 않은 바위가 위치 …에 있음.
.5 저조시에 좌초될 위험이 있음.
.6 본선/MV…은 재부양하기 위하여 화물을 투하하겠음.
.6.1 경고! IMO-등급…의 화물을 투하하지 마시오.

.7 When do you/does MV … expect to refloat?

.7.1 I expect/MV … expects to refloat

~ at … UTC.

~ when tide rises.

~ when weather improves.

~ when draft decreases.

~ with tug assistance/… .

.8 Can you/can MV … beach?[49)]

.8.1 I/MV … can/will beach in position … .

.8.2 I/MV … cannot beach.

.5 List, danger of capsizing

.1 I have/MV … has dangerous list to port/starboard.

.2 I/MV … will

~ transfer cargo/bunkers to stop listing.

~ jettison cargo to stop listing.

.3 I am/MV … in danger of capsizing (- list increasing).

.6 Sinking

.1 I am/MV … sinking after collision/grounding/flooding/explosion/… .

.2 I require/MV … requires assistance.

.3 I am/MV … proceeding to your assistance.

.4 ETA at distress position within … hours/at … UTC.

.7 Disabled and adrift

.1 I am/MV …

~ not under command.

~ adrift.

~ drifting at … knots to …(*cardinal points*).

.7 언제 귀선/MV…이 재부양하리라고 예상하는가?

.7.1 본선/MV…은 세계시…시에 재부양하리라고 예상됨.

본선/MV…은 조석이 상승하면 재부양하리라고 예상됨.

본선/MV…은 기상이 호전되면 재부양하리라고 예상됨.

본선/MV…은 흘수가 감소하면 재부양하리라고 예상됨.

본선/MV…은 예인선의 원조를 받으면/…하면 재부양하리라고 예상됨.

.8 귀선/MV…은 좌안할 수 있는가?

.8.1 본선/MV…은 위치 … 에서 좌안할 수 있음/좌안하겠음.

.8.2 본선/MV…은 좌안할 수 없음.

.5 횡경사, 전복의 위험

.1 본선/MV…은 좌현쪽/우현쪽으로 위험하게 횡경사 되었음.

.2 본선/MV…은 횡경사를 중지시키기 위하여 화물/연료를 이송하겠음.

본선/MV…은 횡경사를 중지시키기 위하여 화물을 투하하겠음.

.3 본선/MV…은 전복의 위험이 있음(횡경사가 증가하고 있음)

.6 침몰

.1 본선/MV…은 충돌/좌초/침수/폭발/…한 후 침몰하고 있음.

.2 본선/MV…은 원조를 필요로 함.

.3 본선/기선…은 귀선을 원조하기 위하여 항진하고 있음.

.4 조난 위치에 …시간 이내에/세계시로…시에 도착할 것으로 예상함.

.7 파손 및 표류

.1 본선/MV…은 운전이 부자유함.

본선/MV…은 표류 중임.

본선/MV…은 … 노트로 …(방점) 방향으로 표류중임.

.2 I require/MV ··· requires tug assistance.

.8 Armed attack/piracy[50)]

.1 I am/MV ··· under attack by pirates.

.1.1 I/MV ··· was under attack by pirates.

.2 I require/MV ··· requires assistance.

.3 What kind of assistance is required?

.3.1 I require/MV ··· requires
- ~ medical assistance.
- ~ navigational assistance.
- ~ military assistance.
- ~ tug assistance.
- ~ escort/··· .

.4 Report damage.

.4.1 I have/MV ··· has
- ~ no damage.
- ~ damage to navigational equipment/··· .

.4.2 I am/MV ··· not under command.

.5 Can you/can MV ··· proceed?

.5.1 Yes, I/MV ··· can proceed.

.5.2 No, I/MV ··· cannot proceed.

.9 Undesignated distress

.1 I have/MV ··· has problems with cargo/engine(s)/navigation /··· .

.2 I require/MV ··· requires ··· .

.10 Abandoning vessel

.1 I/crew of MV ··· must abandon vessel ··· after explosion/collision/grounding/flooding/piracy/armed attack/··· .

.2 본선/MV선…은 예인선의 원조를 필요로 함.

.8 무장 공격/해적

.1 본선/기선…은 해적의 공격을 받고 있음.
.1.1 본선/기선…은 해적의 공격을 받았음.
.2 본선/기선…은 원조를 필요로 함.
.3 어떤 종류의 원조를 필요로 하는가?
.3.1 본선/기선…은 의료상의 원조를 필요로 함.
본선/기선…은 항해상의 원조를 필요로 함.
본선/기선…은 군사상의 원조를 필요로 함.
본선/기선…은 예인선의 원조를 필요로 함.
본선/기선…은 호송/…을 필요로 함.
.4 피해를 보고하시오.
.4.1 본선/기선…은 피해를 입지 않았음.
본선/기선…은 항해장비/…에 피해를 입었음.
.4.2 본선/기선…은 운전이 부자유함.
.5 귀선/기선…은 항진할 수 있는가?
.5.1 예, 본선/기선…은 항진할 수 있음.
.5.2 아니오, 본선/기선…은 항진할 수 없음.

.9 명시되지 않은 조난

.1 본선/기선…은 화물/기관/항해상/…에 문제점을 가지고 있음.
.2 본선/기선…은 …을 요청함.

.10 퇴선

.1 이쪽/기선…의 승무원은 폭발/충돌/좌초/침수/해적/무장 공격/… 이후 퇴선 해야함.

.11 Person overboard[51)]

.1 I have/MV ··· has lost persons overboard in position ··· .
.2 Assist with search in vicinity of position ··· .
.3 All vessels in vicinity of position ··· keep sharp lookout and report to ··· .
.4 I am/MV··· is proceeding for assistance - ETA at ··· UTC/within ··· hours.
.5 Search in vicinity of position ··· .
.5.1 I am/MV ··· is searching in vicinity of position ··· .
.6 Aircraft ETA at ··· UTC/within ··· hours to assist in earch.
.7 Can you continue search?
.7.1 Yes, I can continue search.
.7.2 No, I cannot continue search.
.8 Stop search.
.8.1 Return to ··· .
.8.2 Proceed with your voyage.
.10 What is the result of search?
.10.1 The result of search is negative.
.11 I/MV ··· located/picked up person(s) in position ··· .
.12 Person picked up is crewmember/passenger of MV ··· .
.13 What is condition of person(s)?
.13.1 Condition of person(s) bad/good.
.13.2 Person(s) dead.

AI/1.2 Search and Rescue communication

.1 SAR communications (specifying or supplementary to 1.1)

.1 I require/MV ··· requires assistance.
.2 I am/MV ··· proceeding to your assistance.
.3 What is your MMSI number?
.3.1 My MMSI number is ··· .

.11　사람의 선외 추락

.1　본선/기선…은 위치…에서 사람이 선외로 실종되었음.
.2　위치…부근에서 수색 작업을 원조하시오.
.3　위치…부근의 모든 선박은 엄중한 견시를 유지하고…에 보고하시오.
.4　본선/기선…는 원조하기 위하여 항진하고 있으며- 도착 예정 시각은 세계시…시/… 시간 이내임.
.5　위치…부근을 수색하시오.
.5.1　본선/기선…는 위치…부근을 수색하고 있음.
.6　수색을 원조하기 위한 항공기의 도착 예정 시각은 세계시 …시/…시간 이내임.
.7　그 쪽은 수색을 계속할 수 있는가?
.7.1　예, 이쪽은 수색을 계속할 수 있음.
.7.2　아니오, 이쪽은 수색을 계속할 수 없음.
.8　수색을 중지하시오.
.8.1　… 로 돌아가시오.
.8.2　귀선의 항해(계획)에 따라 항진하시오.
.10　수색의 결과는 어떠한가?
.10.1　수색의 결과는 부정적임.
.11　본선/기선…는 위치…에서 사람(들)을 위치 확인/인양하였음.
.12　인양된 사람은 기선…의 승무원/여객임.
.13　사람의 상태는 어떠한가?
.13.1　사람의 상태는 나쁨/좋음.
.13.2　사람이 사망하였음.

AI/1.2 수색과 구조 통신

.1　SAR 통신(1.1을 상술 또는 보충함)

.1　본선/기선…은 원조를 필요로 함.
.2　본선/기선…은 귀선을 원조하기 위하여 항진하고 있음.
.3　귀선의 MMSI 번호는 무엇인가?
.3.1　본선의 MMSI 번호는 …임.

.4 What is your position?
.4.1 My position … .
.5 What is your present course and speed?
.5.1 My present course … degrees, my speed … knots.
.6 Report number of persons on board.
.6.1 Number of persons on board: … .
.7 Report injured persons.
.7.1 No person injured.
.7.2 Number of injured persons/casualties: … .
.8 Will you abandon vessel?
.8.1 I will not abandon vessel.
.8.2 I will abandon vessel at … UTC.
.9 Is your EPIRB/SART transmitting?[52)]
.9.1 Yes, my EPIRB/SART is transmitting.
.9.2 Yes, my EPIRB/SART is transmitting by mistake.
.10 Did you transmit a DSC distress alert?
.10.1 Yes, I transmitted a DSC alert.
.10.2 Yes, I transmitted a DSC alert by mistake.
.11 How many lifeboats/liferafts (with how many persons) will you launch?
.11.1 I will launch … lifeboats/liferafts (with … persons).
.12 How many persons will stay on board?
.12.1 No person will stay on board.
.12.2 … persons will stay on board.
.13 What is the weather situation in your position?
.13.1 Wind …(*cardinal points*) force Beaufort … .
.13.2 Visibility good/moderate/poor.
.13.3 Smooth/moderate/rough/high Sea slight/moderate/heavy swell …(*cardinal points*).[53)]
.13.4 Current … knots, to …(*cardinal points*).

.4 귀선의 위치는 어디인가?
.4.1 본선의 위치는 …임.
.5 귀선의 현재 침로와 속력은 얼마인가?
.5.1 본선의 현재 침로는 …도, 본선의 속력은 …노트임.
.6 승선하고 있는 사람의 수를 보고하시오.
.6.1 승선중인 사람의 수는: …명임.
.7 부상자에 대하여 보고하시오.
.7.1 부상자 없음.
.7.2 부상자/사상자 수는: …명임.
.8 귀선은 퇴선하겠는가?
.8.1 본신은 퇴선하지 않겠음.
.8.2 본선은 세계시…시에 퇴선하겠음.
.9 귀선의 EPIRB/SART는 발신중인가?
.9.1 예, 이쪽의 EPIRB/SART는 발신중임.
.9.2 예, 이쪽의 EPIRB/SART는 실수로 인하여 발신중임.
.10 그쪽은 DSC 조난 경보를 발신하였는가?
.10.1 예, 이쪽은 DSC 경보를 발신하였음.
.10.2 예, 이쪽은 DSC 경보를 실수로 인하여 발신하였음.
.11 귀선은 몇 척의 구명정/구명벌을(몇 명의 인원을 태워) 진수시키겠는가?
.11.1 본선은 …척의 구명정/구명벌을(…명의 사람을 태워)진수 시키겠음.
.12 몇 명의 사람이 선내에 남을 것인가?
.12.1 아무도 선내에 남지 않을 것임.
.12.2 …명의 사람이 선내에 남을 것임.
.13 귀선의 위치에서의 기상 상태는 어떠한가?
.13.1 풍향 …(방점) 풍력계급 …임.
.13.2 시정은 좋음/보통임/나쁨.
.13.3 파도는 잔잔함/보통임/거침/높음 너울은 약간임/보통임/심함 이고 …(방점)임.
.13.4 해류는 …노트, …(방점)으로 흐르고 있음.

.14 Are there dangers to navigation?

.14.1 No dangers to navigation.

.14.2 Warning! Uncharted rocks/ice/abnormally low tides/mines/… .

.2 Acknowledgement and/or relay of SAR - messages

.1 Received MAYDAY from MV… at … UTC on VHF Channel …/ frequency… .

.2 Vessel in position …

- ~ on fire
- ~ had explosion.
- ~ flooded.
- ~ in collision (with …).
- ~ listing/in danger of capsizing.
- ~ sinking.
- ~ disabled and adrift.
- ~ abandoned/… .

.3 Vessel requires assistance.

.4 Received your MAYDAY.

.4.1 My position … .

.4.2 I/MV … will proceed to your assistance.

.4.3 ETA at distress position within … hours/at … UTC.

.3 Performing/co-ordinating SAR - operations

The questions are normally asked and advice is given by the On-scene Coordinator (OSC). For further information see IAMSAR Manual, London/Montreal, 1998.

.1 I will act as On-scene Co-ordinator.

.1.1 I will show following signals/lights: … .

.2 Can you proceed to distress position?

.14 항해상의 위험이 존재하는가?
.14.1 항해상의 위험이 없음.
.14.2 경고! 해도에 기재되지 않은 바위들/얼음/이상 저조/기뢰/ …이 있음.

.2 수색 구조 통보의 수신 확인 및/또는 중계

.1 세계시 …시에 VHF 채널…/주파수 …로 기선…으로부터 메 이데이를 수신하였음.
.2 위치 …에 있는 선박에 화재 발생.
위치 …에 있는 선박에 폭발 발생.
위치 …에 있는 선박이 침수함.
위치 …에 있는 선박이 (…과) 충돌함.
위치 …에 있는 선박이 횡경사 중임/전복의 위험이 있음.
위치 …에 있는 선박이 가라앉고 있음.
위치 …에 있는 선박이 파손되어 표류 중임.
위치 …에 있는 선박이 유기되었음/… 되었음.
.3 선박이 원조를 필요로 하고 있음.
.4 귀선의 메이데이를 수신하였음.
.4.1 이쪽의 위치는 …임.
.4.2 이쪽/기선…는 귀선을 원조하기 위해 항진하겠음.
.4.3 조난 위치에 도착 예정 시각은 …시간 이내/세계시 …시임.

.3 수색 구조 작업의 수행/조정

통상적으로 현장 지휘자(OSC)에 의하여 질문이 행하여지고, 또한 지시가 내려지게 됨. 더 자세한 사항은 1998년, 런던/몬트리얼, IAMSAR 편람을 참조할 것.

.1 본선은 현장 지휘자의 역할을 하겠음.
.1.1 본선은 다음과 같은 신호/등화를 표시하겠음: … .
.2 귀선은 조난 위치로 항진할 수 있는가?

.2.1 Yes, I can proceed to distress position.
.2.2 No, I cannot proceed to distress position.
.3 What is your ETA at distress position?
.3.1 My ETA at distress position within … hours/at … UTC.
.4 MAYDAY position is not correct.
.4.1 Correct MAYDAY position is … .
.5 Vessels are advised to proceed to position … to start rescue.
.6 Carry out search pattern … starting at … UTC.
.7 Initial course … degrees, search speed … knots.
.8 Carry out radar search.
.9 MV … allocated track number … .
.10 MV/MVs … adjust interval between vessels to … kilometres/ nautical miles.
.11 Adjust track spacing to … kilometres/nautical miles.
.12 Search speed now … knots.
.13 Alter course
~ to … degrees (- at … UTC).
~ for next leg of track now/at … UTC.
.14 We resume search in position … .
.15 Crew has abandoned vessel/MV … .
.16 Keep sharp lookout for lifeboats/liferafts/persons in water/… .

.4 Finishing with SAR - operations

.1 What is the result of search?
.1.1 The result of search is negative.
.2 Sighted
~ vessel in position … .
~ lifeboats/life rafts in position … .
~ persons in water/… in position … .
.3 Continue search in position … .
.4 Can you pick up survivors?

.2.1 예, 본선은 조난 위치로 항진할 수 있음.

.2.2 아니오, 본선은 조난 위치로 항진할 수 없음.

.3 귀선의 조난 위치에의 도착 예정 시각은 언제인가?

.3.1 본선의 조난 위치에의 도착 예정 시각은…시간 이내/세계시 …시임.

.4 메이데이 위치는 올바르지 않음.

.4.1 올바른 메이데이 위치는 …임.

.5 선박들은 구조개시를 위하여 위치 …로 항진을 권고함.

.6 세계시…시부터 수색 방식…을 실시하시오.

.7 시작 침로 …도, 수색 속력 …노트임.

.8 레이더 수색을 실시하시오.

.9 기선 …는 수색 진로 번호…에 할당됨.

.10 기선…/기선…들은 선박 간격을…킬로미터/해리로 조정하시오.

.11 수색 진로 사이의 간격을…킬로미터/해리로 조정하시오.

.12 현재 수색 속력은…노트임.

.13 (세계시…시에)…도로 변침하시오.
지금/세계시…시에 수색 진로의 다음 직행구간으로 변침 하시오.

.14 우리는 위치…에서 수색을 재개함.

.15 승무원이 선박으로부터/기선…으로부터 퇴선하였음.

.16 구명정/구명벌/물속의 사람/…에 대하여 엄중한 견시를 하시오.

.4 수색 구조 작업의 종료

.1 수색의 결과는 어떠한가?

.1.1 수색의 결과는 부정적임.

.2 위치…에 있는 선박을 보았음.
위치…에 있는 구명정/구명벌을 보았음.
위치…에 있는 물속의 사람/…을 보았음.

.3 위치…에서 수색을 계속하시오.

.4 그 쪽은 생존자들을 인양할 수 있는가?

.4.1 Yes, I can pick up survivors.
.4.2 No, I cannot pick up survivors.
.5 MV …/I will proceed to pick up survivors.
.5.1 Stand by lifeboats/liferafts.
.6 Picked up
~ … survivors in position … .
~ … lifeboats/liferafts (with … persons/casualties) in position … .
~ … persons/casualties in lifejackets in position ….
~ … in position … .
.7 Survivors in bad/good condition.
.8 Do you require medical assistance?
.8.1 Yes, I require medical assistance.
.8.2 No, I do not require medical assistance.
.9 Try to obtain information from survivors.
.10 There are
~ still … lifeboats/liferafts with survivors.
~ no more lifeboats/liferafts.
.11 Total number of persons on board was … .
.12 All persons/… persons rescued.
.13 You/MV … may stop search and proceed with voyage.
.14 There is no hope to rescue more persons.
.15 We finish with SAR - operations.

AI/1.3 Requesting medical assistance
.1 I require/MV … requires medical assistance.
.2 What kind of assistance is required?
.2.1 I require/MV … requires
~ boat for hospital transfer.
~ radio medical advice.
~ helicopter with doctor (to pick up person(s)).

.4.1 예, 이쪽은 생존자들을 인양할 수 있음

.4.2 아니오, 이쪽은 생존자들을 인양할 수 없음.

.5 기선…/본선은 생존자들을 인양하러 항진하겠음.

.5.1 구명정/구명벌을 대기하시오.

.6 위치…에서 …명의 생존자들을 인양하시오.

위치…에서(…사람/사상자를 태운)…척의 구명정/구명벌을 인양하시오.

위치…에서 구명동의를 착용한…명의 사람/사상자를 인양하시오.

위치…에서…를 인양하시오.

.7 생존자들은 나쁜/좋은 상태임.

.8 그 쪽은 의료상의 원조를 필요로 하는가?

.8.1 예, 이쪽은 의료상의 원조를 필요로 함.

.8.2 아니오, 이쪽은 의료상의 원조를 필요로 하지 않음.

.9 생존자들로부터 정보를 입수하여보시오.

.10 생존자가 있는 구명정/구명벌은 아직도 … 척이 있음.

구명정/구명벌은 더 이상 없음.

.11 승선하고 있던 사람은 전부…명이었음.

.12 모든 사람/…명의 사람이 구조되었음.

.13 귀선/기선…는 수색을 중지하고 항해(계획)에 따라 항진하여도 됨.

.14 더 이상의 사람을 구조할 가망성은 없음.

.15 우리는 수색 구조 작업을 종료함.

AI/1.3 의료 원조의 요청

.1 본선/기선…은 의료상의 원조를 필요로 함.

.2 어떤 종류의 원조를 필요로 하는가?

.2.1 본선/기선…은 병원 이송을 위한 보우트를 필요로 함.

본선/기선…은 무선 의료 조언을 필요로 함

본선/기선…은 (사람을 태우기 위해)의사가 탄 헬리콥터를 필요로 함

.3 I/MV … will
~ send boat.
~ send helicopter with doctor.
~ send helicopter to pick up person(s).
~ arrange for radio medical advice on VHF channel … /frequency … .

.4 Boat/helicopter ETA at … UTC/within … hours.

.5 Do you have doctor on board?

.5.1 Yes, I have doctor on board.

.5.2 No, I have no doctor on board.

.6 Can you make rendezvous in position … ?

.6.1 Yes, I can make rendezvous in position … at … UTC/within … hours.

.6.2 No, I cannot make rendezvous.

.7 I/MV … will send boat/helicopter to transfer doctor.

.8 Transfer person(s) to my vessel/to MV … by boat/helicopter.

.9 Transfer of person(s) not possible.

AI/2 Urgency traffic

Safety of a vessel (other than distress).

Note: An urgency traffic has always to commence with stating the position of the calling vessel if it is not included in the DSC alert.

.1 Technical failure

.1 I am/MV … not under command.

.2 What problems do you have/does MV … have?

.2.1 I have/MV … has problems with engine(s)/steering gear/propeller/… .

.3 I am/MV … is manoeuvring with difficulty.

.4 Keep clear of me/MV … .

.3 본선/기선…은 보우트를 보내겠음.

본선/기선…은 의사가 탄 헬리콥터를 보내겠음.

본선/기선…은 사람을 태우기 위해 헬리콥터를 보내겠음.

본선/기선…은 VHF 채널…으로/주파수…으로 무선 의료 조언을 준비 하겠음.

.4 보우트/헬리콥터의 도착 예정 시각은 세계시…시/…시간이내임.

.5 귀선은 선내에 의사가 있는가?

.5.1 예, 본선은 선내에 의사가 있음.

.5.2 아니오, 본선은 선내에 의사가 없음.

.6 귀선을 위치…에서 만날(랑데뷰할) 수 있는가?

.6.1 예, 본선은 위치…에서 세계시…시에/…시간 이내에 만날 수 있음.

.6.2 아니오, 본선은 만날 수 없음.

.7 본선/기선…은 의사를 이송하기 위하여 보우트/헬리콥터를 보내겠음.

.8 본선/기선…에 보우트/헬리콥터로 사람을 이송하시오.

.9 사람의 이송은 가능하지 않음.

AI/2 긴급 통화

선박의 안전(조난 이외).

주; 호출하는 선박의 위치가 DSC 경보에 포함되지 않았으면 긴급 통화는 항상 호출하는 선박의 위치를 말하면서 시작해야 한다.

.1 기술상의 결함

.1 본선/기선…은 운전이 부자유함.

.2 귀선/기선…에는 어떠한 문제를 가지고 있는가.

.2.1 본선/기선…에는 주기관/조타기/프로펠러/…에 문제가 있음.

.3 본선/기선…은 어렵게 조종하고 있음.

.4 본선/기선…에서 떨어지시오.

.5 Navigate with caution.
.6 I require/MV … requires tug assistance/escort/… .
.7 I try/MV … tries to proceed without assistance.
.8 Stand by on VHF Channel … /frequency … .
.8.1 Standing by on VHF Channel … /frequency … .

.2 Cargo

.1 I have/MV … has lost dangerous goods of IMO-Class … in position … .
.2 Containers/barrels/drums/bags/… with dangerous goods of IMO -Class … adrift near position … .
.3 I am/MV … is spilling
~ dangerous goods of IMO-Class … in position … .
~ crude oil/… in position … .
.4 I require/MV … requires oil clearance assistance - danger of pollution.
.5 I am/MV … is dangerous source of radiation.

.3 Ice damage

.1 I have/MV … has damage above/below waterline.
.2 What kind of assistance is required?
.2.1 I require/MV … requires
~ tug assistance.
~ ice-breaker assistance/escort/… .
.3 I have/MV … has stability problems - heavy icing.
.4 Can you proceed without assistance?
.4.1 Yes, I can proceed without assistance.
.4.2 No, I cannot proceed without assistance.
.5 Stand by on VHF Channel … /frequency … .
.5.1 Standing by on VHF Channel … /frequency … .

.5 주의해서 항해하시오.
.6 본선/기선…은 예인선의 원조/호송/…를 필요로 함.
.7 본선/기선…은 원조 없이 항진하여보고 있음.
.8 VHF채널…/주파수…로 대기하시오.
.8.1 VHF채널…/주파수…로 대기하고 있음.

.2 화물

.1 본선/기선…은 위치…에서 IMO등급…인 위험 화물을 잃었음.
.2 위치…부근에 IMO 등급…인 위험 화물이 들어있는 컨테이너/통/드럼/자루/…들이 표류하고 있음.
.3 본선/기선…가 위치…에서 IMO 등급…인 위험 화물을 유출하고 있음.
본선/기선…가 위치…에서 원유/…를 유출하고 있음.
.4 본선/기선…이 유류 제거상의 원조를 필요로 함-오염의 위험이 있음.
.5 본선/기선…는 방사성 물질의 위험한 유출원임.

.3 결빙에 의한 손상

.1 본선/기선…은 수선 상부/하부에 손상이 있음.
.2 어떤 종류의 원조가 필요한가?
.2.1 본선/기선…은 예인선의 원조를 필요로 함.
본선/기선…은 쇄빙선의 원조/호송/…를 필요로 함.
.3 본선/기선…는 -심한 착빙으로 인하여- 복원성에 문제가 있음.
.4 귀선은 원조 없이 항진할 수 있는가?
.4.1 예, 본선은 원조 없이 항진할 수 있음.
.4.2 아니오, 본선은 원조 없이 항진할 수 없음.
.5 VHF채널…/주파수…에 대기하시오.
.5.1 VHF채널…/주파수…에 대기하고 있음.

AI/3 Safety Communications

AI/3.1 Meteorological and hydrological conditions

.1 Winds, storms, tropical storms, sea state

.1 What is wind direction and force in your position/in position …?

.1.1 Wind direction …(*cardinal points*), force Beaufort … in my position/in position … .

.2 What wind is expected in my position/in position …?

.2.1 The wind in your position/in position … is expected
~ from direction …(*cardinal points*), force Beaufort ….
~ to increase/decrease.
~ variable.

.3 What is the latest gale/storm warning?

.3.1 The latest gale/storm warning is as follows:
Gale/storm warning. Winds at … UTC in area …(*met. area*)[54] from direction …(*cardinal points*) and force Beaufort … backing/veering to …(*cardinal points*).

.4 What is the latest tropical storm warning?

.4.1 The latest tropical storm warning is as follows:[55]
Tropical storm warning at … UTC. Hurricane …(*name*) /tropical cyclone/tornado/willy-willy/typhoon …(*name*) with central pres- sure of … millibars/hPascals located in position …. Present movement …(*cardinal points*) at … knots. Winds of … knots within radius of … miles of centre. Seas smooth/moderate/rough/high.
Further information on VHF Channel …/frequency … .

.5 What is the atmospheric pressure in your position/in position…?

.5.1 The atmospheric pressure in my position/in position … is … millibars/hPascals.[56]

AI/3 안전 통신

AI/3.1 기상 및 해수의 상황

.1 바람, 폭풍, 열대성 폭풍, 해면 상태

.1 귀선의 위치에서/위치…에서 풍향 및 풍속은 어떠한가?

.1.1 본선의 위치에서/위치…에서 풍향은…(방점), 풍력계급…임.

.2 본선의 위치에서/위치…에서 어떤 바람이 예상되는가?

.2.1 귀선의 위치에서/위치…에서 풍향은…(방점)쪽, 풍력계급은…으로 예상
귀선의 위치에서/위치…에서 바람은 세어질/약해질 것으로 예상됨.
귀선의 위치에서/위치…에서 바람은 가변적일 것으로 예상됨.

.3 최신의 강풍/폭풍 경보는 어떠한가?

.3.1 최신의 강풍/폭풍 경보는 다음과 같음:
강풍/폭풍 경보. 세계시…시, …해역의 풍향은…(방점), 풍력계급은…이며 풍향 반전/순전하여…(방점)방향으로 됨.

.4 최신의 열대성 폭풍 경보는 어떠한가?

.4.1 최신의 열대성 폭풍 경보는 다음과 같음:
세계시…의 열대성 폭풍 경보. 허리케인…(이름)/열대성 사이클론/토네이도/윌리윌리/태풍…(이름)의 중심기압은… 밀리바/헥토파스칼이고 위치는…임.
현재의 이동 방향은…(방점), …노트임. 중심으로부터 반경…해리 이내의 풍속은…노트임. 파도는 잔잔함/보통임/ 거침/높음.
추가의 통보가 VHF 채널…/주파수…에서 있음.

.5 그쪽 위치/위치…에서의 기압은 얼마인가?

.5.1 이쪽 위치/위치…에서의 기압은…밀리바/헥토파스칼임.

.6 What is the barometric change in your position/in position …?

.6.1 The barometric change in my position/in position … is …milli-bars/hPascals per hour/within the last … hours.

.6.2 The barometer is steady/dropping (rapidly)/rising (rapidly).

.7 What maximum winds are expected in the storm area?

.7.1 Maximum winds of … knots are expected

~ in the storm area.

~ within a radius of … kilometres/miles of the centre.

~ in the safe/dangerous semicircle.[57)]

.8 What is sea state in your position/in position … ?

.8.1 The smooth/moderate/rough/high sea slight/moderate/heavy swell in my position/in position … is … metres from…(*cardinal points*).

.9 Is the sea state expected to change (-within the next … hours)?

.9.1 No, the sea state is not expected to change (-within the next … hours).

.9.2 Yes, a sea/swell of … metres from …(*cardinal points*) is expected (- within the next … hours).

.10 A tsunami/an abnormal wave is expected by … UTC.[58)]

.2 Restricted visibility

.1 What is visibility in your position/in position … ?

.1.1 Visibility in my position/in position … is … metres/nautical miles[59)]

.1.2 Visibility is restricted by mist/fog/snow/dust/rain.

.1.3 Visibility is increasing/decreasing/variable.

.2 Is visibility expected to change in my position/in position … (within the next … hours)?

.2.1 No, visibility is not expected to change in your position/in position … (- within the next … hours).

.6 그쪽 위치/위치…에서의 기압 변화는 얼마인가?

.6.1 이쪽 위치/위치…에서의 기압 변화는 시간당/지난…시간 동안 … 밀리바/헥토파스칼임.

.6.2 기압은 변화 없음/(급격히)하강함/(급격히)상승함.

.7 폭풍권 내에서의 최대 풍속은 얼마로 예상되는가?

.7.1 폭풍권 내에서의 최대 풍속은…노트로 예상됨.

중심으로부터 반경 …킬로미터/해리 이내에서 최대 풍속은… 노트로 예상됨.

가항 반원/위험 반원에서의 최대 풍속은…노트로 예상됨.

.8 그쪽의 위치/위치…에서의 해면 상태는 어떠한가?

.8.1 이쪽의 위치/위치…에서, 파도는 잔잔함/보통임/거침/높음 너울은 약간임/보통임/심함이고 높이…미터, …(방점)쪽에서 오고 있음.

.9 (앞으로 … 시간 이내에)해면 상태는 변할 것으로 예상되는가?

.9.1 아니오, (앞으로 … 시간 이내에)해면 상태는 변할 것으로 예상되지 않음

.9.2 예, (앞으로 … 시간 이내에) …미터, …(방점)방향의 파도/너울이 예상됨

.10 세계시 …시까지 쓰나미/비정상적인 파도가 예상됨.

.2 제한된 시정

.1 그 쪽의 위치/위치…에서의 시정은 얼마인가?

.1.1 이쪽의 위치/위치…에서의 시정은…미터/해리임.

.1.2 시정은 엷은 안개/안개/눈/먼지/비에 의하여 제한됨.

.1.3 시정은 증가하고 있음/감소하고 있음/가변적임.

.2 (앞으로 … 시간 이내에)이쪽의 위치/위치…에서의 시정은 변할 것으로 예상되는가?

.2.1 아니오, (앞으로 … 시간 이내에)그 쪽의 위치/위치…에서의 시정은 변할 것으로 예상되지 않음

.2.2 Yes, visibility is expected to increase/decrease to … metres/nautical miles in your position/in position … (within the next … hours).

.2.3 Visibility is expected to be variable between … and … metres/nautical miles in your position/in position … (within the next … hours).

.3 Ice

.1 What is the latest ice information?

.1.1 Ice warning. Ice/iceberg(s) located in position … /reported in area around … .

.1.2 No ice located in position …/reported in area around … .

.2 What ice situation is expected in my position/area around … ?

.2.1 Ice situation is
- ~ not expected to change in your position/area around… .
- ~ expected to improve/deteriorate in your position /area around … .

.2.2 Thickness of ice is expected to increase/decrease in your posi-tion/area around … .

.3 Navigation is dangerous in area around … due to floating ice/pack ice/iceberg(s).[60)]

.4 Navigation in area around … is only possible
- ~ for high-powered vessels of strong construction.
- ~ with ice-breaker assistance.

.5 Area around … temporarily closed for navigation.

.6 Danger of icing in area around … .

.4 Abnormal tides

.1 The present tide is … metres above/below datum in position …

.2 The tide is … metres above/below prediction.

.3 The tide is rising/falling.

.4 Wait until high/low water.

.2.2 예, (앞으로 … 시간 이내에)그 쪽의 위치/위치…에서의 시정은…미터/해리로 증가/감소할 것으로 예상됨

.2.3 (앞으로 … 시간 이내에)그 쪽의 위치/위치…에서의 시정은…와…미터/해리 사이에서 가변적일 것으로 예상됨

.3 얼음

.1 얼음에 대한 최신의 정보는 무엇인가?

.1.1 얼음 경보. 얼음/빙산이 위치…에서 위치확인 됨/…부근 해역에 있다고 보고되었음.

.1.2 위치…에서 얼음이 확인 안됨/…부근 해역에서 얼음이 보고 되지 않았음.

.2 본선의 위치/…부근 해역에서 어떤 얼음 상태가 예상되는가?

.2.1 귀선의 위치/…부근 해역에서 얼음 상태는 변할 것으로 예상 되지 않음.

귀선의 위치/…부근 해역에서 얼음 상태는 좋아질 것으로/ 나빠질 것으로 예상됨.

.2.2 귀선의 위치/…부근 해역에서 얼음의 두께는 증가할 것으로/ 감소할 것으로 예상됨.

.3 부빙/부빙군/빙산으로 인하여…부근 해역에서는 항해가 위험

.4 …부근에서의 항해는 고강도의 고출력선 만이 가능함.

…부근에서의 항해는 쇄빙선의 원조가 있어야만 가능함.

.5 …부근 해역은 일시 항해 금지임.

.6 …부근 해역에 결빙의 위험이 있음.

.4 이상 조석

.1 위치…에서의 현재 조석은 기본수준면 상/하…미터임.

.2 조석은 예보 보다…미터 위임/아래임.

.3 조석은 창조(상승중)임/낙조(하강중)임.

.4 고조/저조 시까지 기다리시오.

.5 Abnormally high/low tides are expected in position … at about … UTC/within … hours.
.6 Is the depth of water sufficient in position … ?
.6.1 Yes, the depth of water is sufficient in position … .
.6.2 No, the depth of water is not sufficient in position … .
.6.3 The depth of water is … metres in position … .
.7 My draft is … metres - can I enter/pass …*(charted name of place)*?
.7.1 Yes, you can enter/pass …*(charted name of place).*
.7.2 No, you cannot enter/pass …(*charted name of place)* - wait until … UTC.
.8 The charted depth of water is increased/decreased by … metres due to sea state/winds.

AI/3.2 Navigational warnings involving

.1 Land- or seamarks

Defects

.1 …(*charted náme of light/buoy*) in position …
~ unlit/unreliable/damaged/destroyed/off station /missing.

Alterations

.2 …(*charted name of light/buoy*) in position …
~ (temporarily) changed to …(*full characteristics*).
~ (temporarily) removed.
~ (temporarily) discontinued.

New and moved

.3 …(*charted name of light/buoy*) …(*full characteristics*)
~ established in position …
~ re-established in position …
~ moved … kilometres/nautical miles in …(*direction*) to position …

.5 위치…에서 세계시…시 경/…시간 이내에 이상 고조/저조 예상됨.
.6 위치…에서 수심은 충분한가?
.6.1 예, 위치…에서 수심은 충분함.
.6.2 아니오, 위치…에서 수심은 충분하지 않음.
.6.3 위치…에서 수심은 …미터임.
.7 본선의 흘수는…미터임- 본선은…(해도에 기재된 지명)를 진입/통과할 수 있는가?
.7.1 예, 귀선은…(해도에 기재된 지명)를 진입/통과할 수 있음.
.7.2 아니오, 귀선은…(해도에 기재된 지명)를 진입/통과할 수 없음. -세계시…시까지 기다리시오.
.8 해면 상태/바람 때문에 해도에 기재된 수심은…미터 증가됨/ 간소됨.

AI/3.2 다음을 포함하는 항행 경보

.1 육상 표지 또는 해상 표지

결함

.1 위치…의 …(등대/부표의 해도상 명칭)은 소등됨/신빙성이 없음/손상됨/파괴됨/제 위치에서 이탈함/없어짐.

변경

.2 위치…의 …(등대/부표의 해도상 명칭)는 …(상세 특성)로 (일시) 변경됨.
위치…의…(등대/부표의 해도상 명칭)는 (일시) 제거됨.
위치…의…(등대/부표의 해도상 명칭)는 (일시) 중지됨

신설 및 이동

.3 …(등대/부표의 해도상 명칭)…(상세특성)이 위치…에 설치됨
…(등대/부표의 해도상 명칭)…(상세특성)이 위치…에 재설치됨
…(등대/부표의 해도상 명칭)이…킬로미터/해리…(방향)쪽으로 이동되어 위치…로 감.

.4 (Note: Only for major fog signal stations.) Fog signal … (*charted name of light/buoy*) in position … inoperative.

.2 Drifting objects

.1 Superbuoy/mine/unlit derelict vessel/ … (*number*) container (s) adrift in vicinity …(*position*) at …(*date and time if known*).

.3 Electronic navigational aids

.1 GPS Satellite …(*number*) unusable from …(*date and time*) to …(*date and time*).[61] Cancel one hour after time of restoration.

.2 LORAN station …(*name or number of master/secondary*) off air from …(*date and time*) to …(*date and time*).[62] Cancel one hour after time of restoration.

.3 RACON …(*name of station*) in position … off air from …(*date and time*) to…(*date and time*).[63]
Cancel one hour after time of restoration.

.4 Seabottom characteristics, wrecks

Use REPORTED when position is unconfirmed, and use LOCATED when position has been confirmed by survey or other means

.1 Uncharted reef/rock/shoal/dangerous wreck/obstruction reported/located in position … .

.2 Dangerous wreck in position… marked by …(*type*) buoy … (*distance in kilometres/nautical miles*) …(*direction*).[64]

.4 (주: 주요 무중신호국 에만)
위치…의 무중신호…(등대/부표의 해도상 명칭)는 작동하지 않음.

.2 표류 물체

.1 …(위치)부근에…(알 수 있을 경우 일, 시)에 대형 부표/기뢰/소등된 유기선/…(개)의 컨테이너가 표류중임.

.3 전자 항행 보조 설비

.1 GPS 위성 …(번)은 …(일시)부터 …(일시)까지 사용불가함.
복구된 시각으로부터 1 시간 지난 후 취소함.

.2 LORAN 국 …(주국/종국의 명칭 또는 번호)은 …(일시) 부터 …(일시)까지 방송 중지됨.
복구된 시각으로부터 1 시간 지난 후 취소함.

.3 위치 …의 RACON …(국명)은 …(일시)부터…(일시)까지 방송 중지됨.
복구된 시각으로부터 1 시간 지난 후 취소함.

.4 해저의 특성, 난파선

위치가 확인 안된 경우는 REPORTED, 측량이나 기타의 방법으로 위치가 확인된 경우는 LOCATED를 쓸 것

.1 위치 …에, 해도에 나와 있지 않은 암초/바위/사주/위험한 난파선/장애물이 보고됨/ 위치 확인됨.

.2 위치 …에 위험한 난파선이 있으며, …(거리로서 킬로미터/ 해리) …(방향)에 …(형식)부표로 표시됨.

.5 Miscellaneous

.5.1 Cable, pipeline and seismic/hydrographic operations

.1 Cable/pipeline operations by …(*vessel*) in vicinity … /along line joining …(*positions*) from …(*date and time*) to …(*date and time*). Wide berth requested (*if requested*). Contact via VHF Channel … (*if requested*).

.2 Seismic survey/hydrographic operations by …(*vessel*) from …(*date and time*) to …(*date and time*) in …(*position*). Wide berth requested. (*if requested*).
Contact via VHF Channel … (*if requested*).

.3 Survey vessel …(*name*) towing …(*length*) seismic cable along line joining/in area bounded by/in vicinity …(*position*) from …(*date and time*) to …(*date and time*).
Wide berth requested (*if requested*).
Contact via VHF Channel … (*if requested*).

.4 Hazardous operations by …(*vessel*) in area bounded by/in vicinity …(*position*) from …(*date and time*) to …(*date and time*). Wide berth requested (*if requested*).
Contact via VHF Channel … (*if requested*).

.5 Current meters/hydrographic instruments moored in…(*position*). Wide berth requested (*if requested*).

.5.2 Diving, towing and dredging operations

.1 Diving/dredging operations by vessel …(*name*) from …(*date and time*) to …(*date and time*) in position … .
Wide berth requested (*if requested*).

.2 Difficult tow from …(*port of departure*)on …(*date*) to …(*destination*) on …(*date*). Wide berth requested.

.5 기타

.5.1 케이블, 파이프라인 및 지진/수로 측량 작업

.1 케이블/파이프라인 작업이 …(일시)부터 …(일시)까지 …(선명)에 의하여 …근처에서/…(위치)에서 만나는 선을 따라 이루어짐. 넓은 여유수면이 요구됨(필요한 경우).
VHF채널 …를 통하여 교신할 것(필요한 경우).

.2 지진 조사/수로 측량 작업이 …(선명)에 의하여 …(일시) 부터 … (일시)까지 위치 …에서 이루어짐.
넓은 여유수면이 요구됨 (필요한 경우)
VHF채널 …를 통하여 교신할 것(필요한 경우)

.3 측량선 …(선명) 이 …(일시)부터 …(일시)까지 …(위치)에서 만나는 선을 따라/…(위치)에 의하여 경계지워진 해역에서/…(위치) 부근에서 …(길이)의 지진 케이블을 예인함.
넓은 여유수면이 요구됨 (필요한 경우)
VHF채널 …를 통하여 교신할 것(필요한 경우)

.4 위험한 작업이 …(선명)에 의하여 …(일시)부터 …(일시) 까지 …(위치)로 경계 지워진 해역에서/ …(위치) 부근에서 행하여짐.
넓은 여유수면이 요구됨(필요한 경우).
VHF채널 …를 통하여 교신할 것.(필요한 경우)

.5 …(위치)에 해류계/수로 측량 기구가 계류됨.
넓은 여유수면이 요구됨.(필요한 경우)

.5.2 잠수, 예인, 준설 작업

.1 선박…(선명)에 의하여 위치 …에서, …(일시)부터 …(일시) 까지 잠수작업/준설 작업이 이루어짐. 넓은 여유수면이 요구됨 (필요한 경우)

.2 …(일자) …(출항지)을 출항하여, …(일자) …(도착지)에 도착하는 어려운 예인 작업이 이루어짐.
넓은 여유수면이 요구됨

.5.3 Tanker transhipment

.1 Transhipment of …(*kind of cargo*) in position … .
Wide berth requested.

.2 I am/MT … spilling oil/chemicals /… in position … .
Wide berth requested.

.3 I am/LNG-tanker/LPG-tanker … leaking gas in position … -avoid passing to leeward.

.4 Oil clearance operations near MT … in position … .
Wide berth requested.

.5.4 Off-shore installations, rig moves

.1 Platform …*(name/number if available)* reported/established in position … at …*(date and time)*.
Wide berth requested *(if requested)*.

.2 Platform …*(name/number if available)* removed from… *(position)* on …*(date)*.

.3 Pipeline/platform …*(name/number if available)* in position … spilling oil/leaking gas. Wide berth requested.

.4 Derelict platform …*(name/number if available)* being removed from …*(position)* at …*(date and time)*. Wide berth requested.

.5.5 Defective locks or bridges

.1 Lock …(*name*) defective.[65]

.1.1 For entering …(*charted name of place*) use lock …(*name*).

.2 Lock/bridge …(*name*) defective.

.2.1 Avoid this area - no possibility for vessels to turn.

.5.6 Military operations

.1 Gunnery/rocket firing/missile/torpedo/underwater ordnance

5.3 탱커의 이적 작업

.1 위치 …에서 …(화물 종류)의 이적 작업이 행하여짐.
넓은 여유수면이 요구됨.

.2 본선/탱커…은 위치 …에서 유류/케미컬/…을 유출하고 있음.
넓은 여유수면이 요구됨

.3 본선/LNG 탱커/LPG탱커 …은 위치 …에서 가스를 누출하고 있음. - 풍하측으로 통과하지 마시오

.4 위치 …에서 탱커…부근의 유류 제거 작업이 행하여짐.
넓은 여유수면이 요구됨

.5.4 해상 구조물, 시추선의 이동

.1 위치 … , …(일시)에 플랫폼 …(알 수 있으면, 명칭/번호)이 보고됨/설치됨. 넓은 여유수면이 요구됨(필요한 경우)

.2 플랫폼 …(알 수 있으면, 명칭/번호)이 ..(위치)에서 ..(날짜)에 제거됨.

.3 위치 …의 파이프라인/플랫폼…(알 수 있으면, 명칭/번호)에서 유류/가스가 새고 있음. 넓은 여유수면이 요구됨.

.4 위치 …에서 유기된 플랫폼…(알 수 있으면, 명칭/번호)이 …(일시)에 제거되고 있음. 넓은 여유수면이 요구됨.

.5.5 갑문 또는 교량의 결함

.1 갑문 …(명칭)은 결함이 있음.

.1.1 …(해도상의 지명)에 진입하려면 갑문 …(명칭)을 사용하시오.

.2 갑문/교량 …(명칭)은 결함이 있음.

.2.1 이 해역을 피하시오-선박이 선회할 수 없음.

.5.6 군사 작전

.1 …(위치들)로 경계 지워진 해역에서 …(일시)부터 …(일시)

exercises in area bounded by …(*positions*) from …(*date and time*) to …(*date and time*). Wide berth requested (*if requested*).

.2 Mine clearing operations from …(*date time*) to …(*date and time*) in area bounded by …(*positions*). Wide berth requested. Contact via VHF channel …(*number*) (*if requested*).

.5.7 Fishery

.1 Small fishing boats in area around … - navigate with caution.

.2 Is fishing gear ahead of me?

.2.1 No fishing gear ahead of you.

.2.2 Yes, fishing gear with buoys/without buoys in position …/area around …- navigate with caution.

.3 Fishing gear has fouled my propeller(s).

.4 You have caught my fishing gear.

.5 Advise you to recover your fishing gear.

.6 Fishing in area … prohibited.

AI/3.3 Environmental protection communications

.1 Located oil spill in position … extending …(*length and width in metres*) to …(*cardinal points*).

.2 Located oil spill

~ in your wake.

~ in the wake of MV … .

.3 I have/MV … has accidental spillage of oil/… .

.4 Can you/MV … stop spillage?

.4.1 Yes, I/MV … can stop spillage.

.4.2 No, I/MV … cannot stop spillage.

.5 What kind of assistance is required?

.5.1 I require/MV … requires

- oil clearance assistance.
- floating booms/oil dispersants/…

까지 사격/로켓발사/미사일/어뢰/해저 무기 훈련이 행하여짐. 넓은 여유수면이 요구됨(필요하면)

.2 …(위치들)로 경계 지워진 해역에서 …(일시)부터 …(일시) 까지 소해작업이 행하여 짐. 넓은 여유수면이 요구됨. VHF채널 …(번호)를 통하여 교신하기 바람(필요하면)

.5.7 어로

.1 소형 어선이 …부근 해역에 있음 - 주의하여 항해하시오.

.2 본선의 전방에 어구가 있는가?

.2.1 귀선의 전방에 어구가 없음.

.2.2 예, 위치 …에/…부근 해역에 부표가 달린/부표가 없는 어구 있음 - 주의하여 항해하시오.

.3 어구가 본선 프로펠러에 감겼음.

.4 귀선이 우리 어구에 걸렸음.

.5 귀선에게 귀선의 어구를 걷어 들이도록 권고함.

.6 …해역에서의 어로 작업은 금지되어 있음.

AI/3.3 환경 보호 통신

.1 위치…에서 유출유를 위치 확인하였으며 …(길이와 폭을 미터로 표시), …(방점)으로 확산 중임.

.2 귀선의 항적에서 유출유를 확인함.
기선…의 항적에서 유출유를 확인함.

.3 본선/기선…은 사고로 유류/…을 유출하였음.

.4 귀선/기선…은 유출을 중지시킬 수 있는가?

.4.1 예, 본선/기선…은 유출을 중지시킬 수 있음.

.4.2 아니오, 본선/기선…은 유출을 중지시킬 수 없음.

.5 어떤 종류의 원조를 필요로 하는가?

.5.1 본선/기선…은 유류 제거상의 원조를 필요로 함.
본선/기선…은 부유 붐/유류 분산제/…를 필요로 함.

.5.2 Stay in vicinity of pollution and co-operate with oil clearance team.

.6 …(*number*) barrels/drums/containers with IMDG - Code marks reported adrift near position … .

.7 Located a vessel dumping chemicals/waste/… in position … .

.7.1 Located a vessel incinerating chemicals/waste/… in position … .

.8 Can you identify the polluter?

.8.1 Yes, I can identify the polluter - polluter is MV … .

.8.2 No, I cannot identify the polluter.

.9 What is course and speed of the polluter?

.9.1 Course of the polluter … degrees, speed … knots.

.9.2 The polluter left the scene.

AI/4 Pilotage

AI/4.1 Pilot request

See AI/6 - .4.3 "Pilot request "

AI/4.2 Embarking/disembarking pilot[66]

.1 Stand by pilot ladder.

.2 Rig the pilot ladder on port side/starboard side … metres above water.

.3 The pilot ladder is rigged on port side/starboard side.

.4 You must rig another pilot ladder

.5 The pilot ladder is unsafe.

.6 What is wrong with the pilot ladder?

.6.1 The pilot ladder

~ has broken/loose steps.

~ has broken spreaders.

~ has spreaders too short.

.5.2 오염의 근처에 머물러 유류 제거 팀과 협조하시오.

.6 위치…부근에 …(숫자)개의 IMDG코드 마크가 있는 통/드럼/컨테이너가 표류 중이라고 보고됨.

.7 위치 …에서 화학약품/폐기물/…를 투기하고 있는 선박을 확인함.

.7.1 위치 …에서 화학약품/폐기물/…를 소각하고 있는 선박을 확인함.

.8 귀선은 오염원 선박을 식별할 수 있는가?

.8.1 예, 본선은 오염원 선박을 식별할 수 있음 -오염원 선박은..임.

.8.2 아니오, 본선은 오염원 선박을 식별할 수 없음.

.9 오염원 선박의 침로와 속력은 얼마인가?

.9.1 오염원 선박의 침로는 …도, 속력은…노트임.

.9.2 오염원 선박은 시야에서 사라졌음.

AI/4 도선 업무

AI/4.1 도선사 요청

AI/6-.4.3 "도선사 요청" 참조

AI/4.2 도선사의 승선/하선

.1 도선사용 사다리를 대기하시오.

.2 좌/우현 수면 상 …미터에 도선사용 사다리를 장착 (설치) 하시오.

.3 도선사용 사다리는 좌/우현 측에 장착(설치)되어 있음.

.4 귀선은 다른 도선사용 사다리를 장착해야 함.

.5 도선사용 사다리는 안전하지 않음.

.6 도선사용 사다리에 무엇이 잘못되어 있는가?

.6.1 도선사용 사다리의 발판이 부서졌음/느슨해 졌음.
도선사용 사다리의 스프레이더가 부서졌음.
도선사용 사다리의 스프레이더가 너무 짧음.

~ is too far aft/forward.

.7 Move the pilot ladder

~ … metres aft/forward.

~ clear of discharge.

.8 Rig the accommodation ladder in combination with the pilot ladder.

.9 Rig the pilot ladder alongside hoist.

.10 Put lights on at the pilot ladder.

.11 Man ropes are required/not required.[67)]

.12 Have a heaving line ready at the pilot ladder.

.13 Correct the list of the vessel.

.14 Make a lee on your port side/starboard side.

.15 Steer … degrees to make a lee.

.16 Keep the sea on your port quarter/starboard quarter.

.17 Make a boarding speed of … knots.

.18 Stop engine(s) until pilot boat is clear.

.19 Put helm hard to port/starboard.

.20 Alter course to …*(cardinal points)* - the pilot boat cannot clear the vessel.

.21 Put engine(s) ahead/astern.

.22 Embarkation is not possible.

.22.1 Boarding arrangements do not comply with SOLAS-Regulations.[68)]

.22.2 Vessel is not suited for the pilot ladder.

AI/4.3 Tug request

.1 Must I take tug(s)?

.1.1 Yes, you must take … tug(s).

.1.2 No, you need not take tug(s).

.2 How many tugs must I take?

.2.1 You must take … tug(s) according to Port Regulations.[69)]

.2.2 You must take … tug(s) fore and … tug(s) aft.

도선사용 사다리가 너무 뒤쪽/앞쪽에 있음.

.7 도선사용 사다리를 뒤쪽/앞쪽으로 …미터 이동하시오.

도선사용 사다리를 배출구에서 피하도록 이동하시오.

.8 현문 사다리를 도선사용 사다리와 조합하여 장착하시오.

.9 도선사용 사다리를 호이스트와 나란히 장착하시오.

.10 도선사용 사다리에 등화를 켜시오.

.11 손잡이 밧줄이 필요함/필요치 않음.

.12 도선사용 사다리에 히빙라인을 준비하시오.

.13 선박의 횡경사를 바로 잡으시오.

.14 귀선의 좌현/우현을 풍하 측으로 하시오.

.15 풍하 측을 만들기 위하여 …도로 조타하시오.

.16 귀선의 좌현/우현 선미에 파도가 오도록 하시오.

.17 승선 속력 … 노트로 만드시오.

.18 도선선이 벗어날 때까지 주기관을 정지하시오.

.19 키를 좌현/우현으로 완전히 돌리시오.

.20 …(방점)으로 변침 하시오-도선선이 본선으로부터 벗어날 수 없음.

.21 주 기관을 전진/후진으로 하시오.

.22 승선이 불가능함.

.22.1 승선 설비가 SOLAS 규정에 따르지 않음.

.22.2 선박이 도선사용 사다리에 맞지 않음.

AI/4.3 예인선 요청

.1 본선은 예인선을 사용해야 하는가?

.1.1 예, 귀선은…척의 예인선을 사용해야 함.

.1.2 아니오, 귀선은 예인선을 사용할 필요가 없음.

.2 본선은 몇 척의 예인선을 사용해야 하는가?

.2.1 항구 규정에 따라서 귀선은 …척의 예인선을 사용해야 함.

.2.2 귀선은 선수에…척, 선미에…척의 예인선을 사용해야 함.

.3 I require … tug(s).
.4 In what position will the tug(s) meet me?
.4.1 The tug(s) will meet you in position … at … UTC.
.4.2 Wait for the tug(s) in position … .
.5 Must I use the towing lines of my vessel?
.5.1 Yes, you must use the towing lines of your vessel.
.5.2 No, you must use the towing lines of the tug.

AI/5 Specials

AI/5.1 Helicopter operations

(H: = from helicopter V: = from vessel)

.1 V: I require a helicopter.
~ to pick up persons.
~ with doctor.
~ with liferaft/… .
.1.1 MRCC: I will send a helicopter with … .
.2 H: MV … , I will drop … .
.3 H: MV … , are you ready for the helicopter?
.3.1 V: Yes, I am ready for the helicopter.
.3.2 V: No, I am not ready for the helicopter (yet).
.3.3 V: Ready for the helicopter in … minutes.
.4 H: MV … , helicopter is on the way to you.
.5 H: MV … , what is your position.
.5.1 V: My position is … .
.6 H: MV … , what is your present course and speed.
.6.1 V: My present course is … degrees, speed is … knots.
.7 H: MV … , make identification signals.
.8 V: I am making identification signals by smoke (buoy)/ search light /flags/signalling lamp/… .
.9 H: MV … , you are identified.
.10 H: MV … , what is the relative wind direction in degrees and knots.

.3 본선은…척의 예인선을 필요로 함.
.4 예인선은 어느 위치에서 본선을 만나게 되는가?
.4.1 예인선은 위치…에서 세계시…시에 귀선과 만나게 됨.
.4.2 위치…에서 예인선을 기다리시오.
.5 본선은 본선의 예인줄을 사용해야 하는가?
.5.1 예, 귀선은 귀선의 예인줄을 사용해야 함.
.5.2 아니오, 귀선은 예인선의 예인줄을 사용해야 함

AI/5 특수 사항

AI/5.1 헬리콥터 작업

(H:=헬리콥터로부터 V:=선박으로부터)

.1 V: 본선은 사람을 태우기 위하여 헬리콥터를 필요로 함.
V: 본선은 의사를 동반한 헬리콥터를 필요로 함.
V: 본선은 구명벌/…을 구비한 헬리콥터를 필요로 함.
.1.1 MRCC: 이쪽은 …를 구비한 헬리콥터를 보내겠음.
.2 H: 기선…, 본기는…을 강하하겠음.
.3 H: 기선…, 귀선은 헬리콥터를 맞이할 준비가 되었는가?
.3.1 V: 예, 본선은 헬리콥터를 맞이할 준비가 되었음.
.3.2 V: 아니오, 본선은 (아직)헬리콥터를 맞이할 준비가 안되었음.
.3.3 V: …분 후에 헬리콥터를 맞이할 준비가 됨.
.4 H: 기선…, 헬리콥터가 귀선에 가고 있는 중임.
.5 H: 기선…, 귀선의 위치는 어디인가?
.5.1 V: 본선의 위치는…임.
.6 H: 기선…, 귀선의 현재의 침로와 속력은 얼마인가?
.6.1 V: 본선의 현재의 침로는 …도 , 속력은…노트임.
.7 H: 기선… , 식별 신호를 하시오.
.8 V: 본선은 발연(부표)/탐조등/신호기/발광 신호등/…에 의한 식별신호를 하고 있음.
.9 H: 기선… , 귀선은 식별되었음.
.10 H: 기선… , 상대 풍향은 몇 도 방향, 몇 노트인가?

.10.1 V: The relative wind direction is … degrees and … knots.
.11 H: MV … , keep the wind on port/starboard bow.
.12 H: MV … , keep the wind on port/starboard quarter.
.13 H: MV … , indicate the landing/pick-up area.[70)]
.13.1 V: The landing/pick-up area is … .
.14 H: MV … , can I land on deck?
.14.1 V: Yes, you can land on deck.
.14.2 V: No, you cannot land on deck (yet).
.14.3 V: You can land on deck in … minutes.
.15 H: MV … , I will use hoist/rescue sling/rescue basket/ rescue net /rescue litter/rescue seat/double lift.
.16 V: I am ready to receive you.
.17 H: MV … , I am landing.
.18 H: MV … , I am starting operation.
.19 H: MV … , do not fix the hoist cable.
.20 H: MV … , operation finished.
.21 H: MV … , I am taking off.

AI/5.2 Ice-breaker operations

.1 Ice-breaker request

.1 I am/MV … is fast in ice in position … .
.2 I require/MV … requires ice-breaker assistance to reach….
.3 Ice-breaker assistance
- ~ will arrive at … UTC/within … hours.
- ~ is not available until … UTC.
- ~ is available only up to latitude … longitude … .
- ~ is suspended until …*(date and time).*
- ~ is suspended after sunset.
- ~ is suspended until favourable weather conditions.
- ~ will be resumed at … UTC.

.10.1 V: 상대 풍향은 …도 방향, …노트임.
.11 H: 기선… , 바람을 좌현/우현 선수에서 받으시오.
.12 H: 기선… , 바람을 좌현/우현 선미에서 받으시오.
.13 H: 기선… , 착륙/인양 장소를 표시하시오.
.13.1 V: 착륙/인양 장소는…임.
.14 H: 기선… , 본기는 갑판에 착륙해도 되는가?
.14.1 V: 예, 귀기는 갑판에 착륙하여도 됨.
.14.2 V: 아니오, 귀기는 (아직)갑판에 착륙하면 안됨.
.14.3 V: 귀기는 …분 후에 갑판에 착륙하여도 됨.
.15 H: 기선… , 본기는 호이스트/구조 슬링/구조 바스켓/구조 네트 /구조들것/구조 의자/ 이중 리프트를 사용하겠음.
.16 V: 본선은 귀기를 받아들일 준비가 되었음.
.17 H: 기선… , 본기는 착륙하고 있음.
.18 H: 기선… , 본기는 작업을 시작함.
.19 H: 기선… , 호이스트의 밧줄을 고정시키지 마시오.
.20 H: 기선… , 작업이 종료되었음.
.21 H: 기선… , 본기는 이륙함.

AI/5.2 쇄빙선 작업

.1 쇄빙선 요청

.1 본선/기선…은 위치…에서 얼음 속에 고착되었음.
.2 본선/기선…은…에 도착하기 위하여 쇄빙선 원조가 필요함.
.3 쇄빙선은 세계시…시에/…시간 이내에 원조하기 위하여 도착 할 것임.
쇄빙선의 원조는 세계시…시까지는 이용 불가능함.
쇄빙선의 원조는 위도… 경도…까지만 이용 가능함.
쇄빙선의 원조는…(일시)까지 일시 중지함.
쇄빙선의 원조는 일몰 이후에는 일시 중지함.
쇄빙선의 원조는 기상이 양호해질 때까지 일시 중지함.
쇄빙선의 원조는 세계시…시에 재개됨.

.2 Ice-breaker assistance for convoy

Ice-breaker commands applying to all the vessels in a convoy have to be immediately confirmed consecutively by each vessel in turn and executed according to the pattern given in GENERAL 4.6. Ice-breaker commands applying to a single vessel are confirmed and executed only by that vessel, this applies also for close coupled towing. When being assisted by an ice-breaker it is important to maintain a continuous listening watch on the appropriate VHF Channel and to maintain a proper lookout for sound and visual signals.

.1 Ice breaker assistance for convoy will start now/at … UTC.
.2 Your place in convoy is number … .
.3 MV … will follow you.
.4 You will follow MV … .
.5 Go ahead and follow me.
.5.1 Do not follow me.
.6 Proceed along the ice channel.
.7 Increase/reduce your speed.
.8 Reverse your engines.
.9 Stop engines.
.10 Keep a distance of … metres/cables between vessels.
.11 Increase/reduce the distance between vessels to … metres/cables.
.12 Stand by for receiving towing line.
.12.1 Stand by for letting go towing line.
.13 Switch on the bow/stern search light[71)]
.14 Stop in present position.
.15 Ice-breaker … will escort you.
.16 Ice-breaker assistance for convoy finished.
.16.1 Open water/light ice conditions ahead.
.17 Proceed by yourself (to area …).

.2 호송 선단에 대한 쇄빙선의 원조

호송 선단 내의 전 선박에 적용되는 쇄빙선의 명령에는 호송 선단 내의 각 선박이 즉시, 연속하여 차례로 수신 확인을 해야 하며, 일반 사항 4.6에 따라 수행하여야 한다. 어떤 한 척의 선박에만 적용되는 쇄빙선의 명령은 해당 선박만 명령의 수신 확인을 하고 수행하면 되며, 이는 밀접 결합 예인의 경우에도 적용된다. 쇄빙선의 원조를 받고 있는 경우에는, 적절한 VHF 채널로 계속 청취 당직을 유지하고 음향 및 시각 신호에 대비하여 적절한 경계를 유지하는 것이 중요하다.

.1 호송 선단에 대한 쇄빙선의 원조는 지금/세계시…시에 시작될 것임.
.2 호송 선단 내에서 귀선의 위치는 …번째임.
.3 기선…이 귀선의 뒤를 따를 것임.
.4 귀선은 기선…뒤를 따를 것임.
.5 전진하시오 그리고 본선을 뒤따르시오.
.5.1 본선의 뒤를 따르지 마시오.
.6 빙간 수로를 따라 항진하시오.
.7 귀선의 속도를 증가/감소시키시오.
.8 귀선은 기관을 역전하시오.
.9 기관을 정지하시오.
.10 선박 사이의 거리를…미터/케이블로 유지하시오.
.11 선박 사이의 거리를…미터/케이블로 증가/감소시키시오.
.12 예인줄을 받을 준비를 하시오.
.12.1 예인줄을 풀어줄 준비를 하시오.
.13 선수/선미의 탐조등을 켜시오.
.14 현재의 위치에 정지하시오.
.15 쇄빙선 …가 귀선을 호송할 것임.
.16 호송 선단에 대한 쇄빙선의 원조는 종료되었음.
.16.1 전방에는 얼음이 없는/엷게 얼어있는 상태임.
.17 (…해역까지)자력으로 항진하시오.

.3 Ice-breaker assistance in close-coupled towing

.1 Stand by for close coupled towing.
.2 Slack out your anchors under the hawse-pipes.
.3 Pass heaving lines through the hawse-pipes.
.4 Receive towing line on deck.
.5 Lash together the eyes of the towing line with manila lashing.
.6 Fasten towing line on your bitts.
.7 I start to draw your bow into the stern notch of the ice-breaker.
.8 Stand by for cutting the manila lashing if required.
.9 Keep yourself in the centre-plane of the ice-breaker.

AI/6 Vessel Traffic Service (VTS) Standard Phrases

Application of Message Markers

In order to especially facilitate shore-to-ship and ship-to-shore communication or when one of the Standard Marine Communication Phrases will not fit the meaning desired, one of the following eight message markers may be used to increase the probability of the purpose of the message being properly understood. It is at the discretion of the shore personnel or the ship's officer whether to use one of the message markers and if so which of them to apply depending on the user`s qualified assessment of the situation. If used the message marker is to be spoken preceding the message or the corresponding part of the message. The IMO VTS Guidelines recommend that in any message directed to a vessel it should be clear whether the message contains **information, advice, warning,** or **instruction** and IMO Standard Marine Communication Phrases should be used where practicable.

.3 밀접 결합 예인에 의한 쇄빙선의 원조

.1 밀접 결합 예인의 준비를 하시오.

.2 귀선의 닻을 호스파이프 아래로 풀어주시오.

.3 히빙라인을 호스파이프로 통과시키시오.

.4 예인줄을 갑판상에 받으시오.

.5 예인줄의 고리들을 마닐라로프로 서로 결박하시오.

.6 예인줄을 귀선의 계주에 고정하시오.

.7 이쪽은 귀선의 선수를 쇄빙선의 선미 홈에 끌어들이기 시작함.

.8 필요하면 마닐라로프의 결박을 절단할 준비를 하시오.

.9 귀선을 쇄빙선의 중앙 종단면상에 위치시키시오.

AI/6 선박 통항 관제 업무(VTS)의 표준 용어

통보 부호의 적용

육상국과 선박 사이의 통신을 특히 수월하게 하기 위하여, 또는 표준 해사 통신 용어 중 하나가 전달하고자 하는 의도와 일치하지 않는 경우에, 전달문이 의도하는 바가 올바르게 이해될 수 있는 가능성을 높이기 위하여 다음 여덟 가지의 통보 부호 중의 하나를 사용할 수 있다. 통보 부호 중의 어느 하나를 사용할 것인지의 여부는 육상 요원 또는 선박의 사관이 판단할 사항이며, 사용할 경우에 어떤 통보 부호를 사용할 것인지는 당시 상황에 대한 담당자의 적절한 평가에 따른다.

통보 부호가 사용된다면 통보 사항에 앞서서, 또는 통보 사항 중의 해당 부분에 앞서서 통보 부호를 말해야 한다. IMO의 VTS 지침에 의하면 어떤 통보 사항이 선박에 통보될 때 그 전달 사항 내에 **통보, 권고, 경고, 지시 사항**이 포함되어 있는지의 여부를 확실히 해야 하며, 실행 가능한 한 IMO 표준 해사 통신 용어를 사용해야 한다.

For further standardized VTS communications, also see other sections of PART AI. For VTS Standard Reporting Procedures see IMO Resolution A.851(20) on "General Principles for Ship Reporting Systems and Ship Reporting Requirements, including guidelines for reporting incidents involving dangerous goods, harmful substances and/or marine pollutants".[72)]

Note: All of the following phrases must come as the culmination (mes-sage content)of a radio message exchange between stations covered by the ITU Radio Regulations, and the relevant calling procedures have to be observed.

Message Markers

(i) INSTRUCTION

This indicates that the following message implies the intention of the sender to influence others by a Regulation.

Comment: This means that the sender, e.g. a VTS - Station or a naval vessel, must have the full authority to send such a message. The recipient has to follow this legally binding message unless s/he has contradictory safety reasons which then have to be reported to the sender.

Example: "INSTRUCTION. Do not cross the fairway."

(ii) ADVICE

This indicates that the following message implies the intention of the sender to influence others by a Recommendation.

Comment: The decision whether to follow the ADVICE still stays with the recipient. ADVICE does not necessarily have to be followed but should be considered very carefully.

표준화된 VTS 교신에 관한 더 자세한 사항은 AI편의 다른 부분들을 참고할 것. VTS의 표준 보고 절차에 관하여는 “IMO 결의서 A.851(20), 위험 화물, 유해 물질 및/또는 해양 오염 물질과 관련된 사건의 보고에 대한 지침이 포함된 선박 보고 제도 및 선박 보고 요구 사항에 관한 일반 원칙”을 참고할 것.

주: 다음의 모든 어귀들은 ITU 전파법규에 정하여진 무선국 사이의 무선 교신 내용의 완결(전달 내용)로써 사용되어야 한다. 그리고, 적절한 호출 절차를 준수하여야 한다.

통보 부호

(i) 지시(INSTRUCTION)

이는, 다른 자에게 법규에 의한 영향을 미치고자 하는 송신자의 의도가 뒤따르는 통보 사항에 포함되어 있음을 나타낸다.

주석: 이는, VTS 국이나 해군 함정과 같은 송신자가 그러한 통보 사항을 발할 수 있는 전적인 권한을 보유하고 있음을 의미한다. 수신자는 이러한 통보 사항에 상반되는 안전상의 이유가 있으면 송신자에게 보고해야 하며 이와 같은 경우가 아니면 법적인 구속력을 갖는 이러한 통보 사항을 준수해야 한다.

보기: “INSTRUCTION. 항로를 횡단하지 마시오.”

(ii) 권고(ADVICE)

이는 뒤따르는 통보 사항에, 다른 자에게 권고에 의한 영향을 미치고자 하는 송신자의 의도가 포함되어 있음을 나타냄.

주석: 권고사항에 따를지를 결정하는 것은 수신자가 결정할 문제이다. 권고사항을 반드시 따라야 되는 것은 아니고 이를 주의 깊게 고려해야 한다.

Example: "ADVICE. (Advise you) stand by on VHF Channel six nine."

(iii) WARNING

This indicates that the following message implies the intention of the sender to inform others about danger.

Comment: This means that any recipient of a WARNING should pay immediate attention to the danger mentioned. Consequences of a ignoring WARNING will be up to the recipient.

Example: "WARNING. Obstruction in the fairway."

(iv) INFORMATION

This indicates that the following message is restricted to observed facts, situations, etc.

Comment: This marker is preferably used for navigational and traffic information, etc.. Consequences of ignoring INFORMATION will be up to the recipient.

Example: "INFORMATION. MV Noname will overtake to the West of you."

(v) QUESTION

This indicates that the following message is of interrogative character.

Comment: The use of this marker removes any doubt on whether a question is being asked or statement being made, especially when interrogatives such as What, Where, Why, Who, How are additionally used at the beginning of the question.

The recipient is expected to return an answer.

Example: "QUESTION.(What is) your present maximum draft?"

보기: "ADVICE. VHF 채널 69에 대기하시오(대기하기를 권고함)"

(iii) 경고(WARNING)

이는 뒤따르는 통보 사항에, 다른 자에게 위험을 알려주고자 하는 송신자의 의도가 포함되어 있음을 나타낸다.

주석: 경고를 수신한 자는 언급된 위험에 즉시 주의를 기울여야 함을 의미한다. 경고를 무시함에 따른 결과는 수신자에게 있다.

보기: "WARNING. 항로 내에 장애물이 존재함"

(iv) 정보(INFORMATION)

이는 뒤따르는 통보 사항이 관찰된 사실이나 상황 등에 국한됨을 나타냄.

주석: 이 정보부호는 주로 항해에 관한 정보나 통항상황에 관한 것으로, 이 정보를 무시함에 따른 결과는 수신자에게 있다.

보기: "INFORMATION. 기선 Noname 이 귀선의 서쪽을 통하여 귀선을 추월할 것임"

(v) 질문(QUESTION)

이는 뒤따르는 통보 사항이 의문을 나타내는 성격의 것임을 나타냄.

주석: 이 통보 부호를 씀으로서, 특히 질문의 첫머리에 What, Where, Why, How, Who와 같은 의문사가 추가적으로 사용되는 경우에, 진술이 이루어지고 있는지, 질문이 이루어지고 있는지에 대한 의혹이 없어진다. 수신자는 응답을 보낼 것으로 기대된다.

보기: "QUESTION. 귀선의 흘수는(얼마인가)?"

(vi) ANSWER

This indicates that the following message is the reply to a previous question.

Comment: Note that an answer should not contain another question.

Example: "ANSWER. My present maximum draft is zero seven metres."

(vii) REQUEST

This indicates that the following message is asking for action from others with respect to the vessel.

Comment: The use of this marker is to signal: I want something to be arranged or provided, e.g. ship's stores requirements, tugs, permission, etc..

Note: REQUEST must not be used involving navigation, or to modify COLREGS.[73)]

Example: "REQUEST. I require two tugs."

(viii) INTENTION

This indicates that the following message informs others about imme- diate navigational action intended to be taken.

Comment: The use of this message marker is logically restricted to messages announcing navigational actions by the vessel sending this message.

Example: "INTENTION. I will reduce my speed."

(vi) 응답(ANSWER)

이는 뒤따르는 통보 사항이 앞의 질문에 대한 대답임을 나타낸다.

주석: 응답 속에 다른 질문 사항을 포함해서는 안 됨을 주목할 것.

보기: "ANSWER. 본선의 현재 최대 흘수는 7 미터임"

(vii) 요구(REQUEST)

이는 뒤따르는 통보 사항에 의해, 다른 당사자가 해당 선박으로 하여금 행동을 취할 것을 요구함을 나타냄.

주석: 이 통보 부호를 사용하는 것은 다음 사항을 표시하는 것임: 본선은 선용품 주문, 예인선, 허가 등 무엇인가가 준비되거나 공급되기를 원한다.

주: REQUEST는 항해에 관하여, 또는 충돌 예방 규칙을 수정하는 데 사용하지 말것.

보기: "REQUEST. 본선은 예인선 두 척을 요청함"

(viii) 의도(INTENTION)

이는 뒤따르는 통보사항에 의하여, 취하고자하는 즉각적인 항해상의 행동을 다른 자에게 알리고자함을 나타냄.

주석: 이 통보 부호의 사용은, 이 통보 사항을 발신하는 선박이 취할 항해상의 행동을 공시하는 통보 내용에 국한된다.

보기: "INTENTION", 본선은 감속하겠음"

AI/6.1 Phrases for acquiring and providing data for a traffic image

.1 Acquiring and providing routine traffic data

.1 What is the name of your vessel and call sign/identification?

.1.1 The name of my vessel is … , call sign …/identification … .

.1.2 Spell the name of your vessel.

.2 What is your flag state?[74)]

.2.1 My flag state is … .

.3 What is your position?

.3.1 My position is … .

.4 What is your present course and speed?

.4.1 My present course is … degrees, my speed is … knots.

.5 From what direction are you approaching?

.5.1 I am approaching from … .

.6 What is your port of destination/destination?

.6.1 My port of destination/destination is … .

.7 What was your last port of call?

.7.1 My last port of call was … .

.8 What is your ETA in position … ?

.8.1 My ETA is … UTC.

.9 What is your ETD from … ?

.9.1 My ETD from … is … UTC.

.10 What is your draft forward/aft?

.10.1 My draft forward/aft is … metres.

.11 What is your present maximum draft ?

.11.1 My present maximum draft is … metres.

.12 What is your freeboard?

.12.1 My freeboard is … metres.

.13 What is your air draft?

.13.1 My air draft is … metres.

AI/6.1 Traffic image 데이터를 획득하고 제공하기 위한 용어

.1 일상적인 통항 정보의 획득과 제공

.1 귀선의 선명, 호출 부호/식별 부호는 무엇인가?
.1.1 본선의 선명은… , 호출 부호는…/식별 부호는…임.
.1.2 귀선 선명의 철자를 말하시오.
.2 귀선의 선적국은 어디인가?
.2.1 본선의 선적국은…임.
.3 귀선의 위치는 어디인가?
.3.1 본선의 위치는…임.
.4 귀선의 현재 침로와 속력은?
.4.1 본선의 현재 침로는…도, 본선의 속도는…노트임.
.5 귀선은 어느 방향에서 접근중인가?
.5.1 본선은…에서 접근중임.
.6 귀선의 목적항/목적지는 어디인가?
.6.1 본선의 목적항/목적지는…임
.7 귀선의 최후 기항지는 어디인가?
.7.1 본선의 최후 기항지는…임
.8 귀선의 위치…에의 도착 예정 시각은?
.8.1 본선의 도착 예정 시각은 세계시…임.
.9 귀선의…로부터의 출발 예정 시각은?
.9.1 본선의 …로부터의 출발 예정 시각은 세계시…임.
.10 귀선의 선수/선미 흘수는 얼마인가?
.10.1 본선의 선수/선미 흘수는…미터임.
.11 귀선의 현재의 최대 흘수는 얼마인가?
.11.1 본선의 현재의 최대 흘수는 …미터임.
.12 귀선의 건현은 얼마인가?
.12.1 본선의 건현은 …미터임.
.13 귀선의 높이는 얼마인가?
.13.1 본선의 높이는…미터임.

.14 Are you underway?
.14.1 Yes, I am underway.
.14.2 No, I am not underway.
.14.3 I am ready to get underway.
.15 What is your full speed/full manoeuvring speed?
.15.1 My full speed/full manoeuvring speed is … knots.
.16 What is your cargo?
.16.1 My cargo is … .
.17 Do you carry any dangerous goods?
.17.1 Yes, I carry the following dangerous goods: … kilogrammes /tonnes IMO Class … .
.17.2 No, I do not carry any dangerous goods.
.18 Do you have any deficiencies/restrictions?
.18.1 No, I have no deficiencies/restrictions.
.18.2 Yes, I have the following deficiencies/restrictions: … .
.19 I am/MV … is constrained by draft.
.20 The maximum permitted draft is … metres.
.21 Do you have any list?
.21.1 Yes, I have a list to port/starboard of … degrees.
.21.2 No, I have no list.
.22 Are you on even keel?
.22.1 Yes, I am on even keel.
.22.2 No, I am trimmed by the head/stern.

.2 Acquiring and providing distress traffic data

See AI/1.1 "Distress communications"

AI/6.2 Phrases for providing VTS services

.1 Information service

These phrases are normally transmitted from the shore.

.1.1 Navigational warnings

.1 Unknown object(s) in position … .

.14 귀선은 항해중인가?
.14.1 예, 본선은 항해중임.
.14.2 아니오, 본선은 항해중이 아님.
.14.3 본선은 항해 개시 준비가 되어있음.
.15 귀선의 전속력/조종 전속력은 얼마인가?
.15.1 본선의 전속력/조종 전속력은…노트임.
.16 귀선의 화물은 무엇인가?
.16.1 본선의 화물은…임.
.17 귀선은 위험 화물을 적재하고 있는가?
.17.1 예, 본선은 다음의 위험 화물을 적재하고 있음: …킬로그램/톤, IMO등급… .
.17.2 아니오, 본선은 위험 화물을 적재하고 있지 않음.
.18 귀선은 어떤 결함/제한 사항을 가지고 있는가?
.18.1 아니오, 본선은 결함/제한 사항을 가지고 있지 않음.
.18.2 예, 본선은 다음의 결함/제한 사항을 가지고 있음:… .
.19 본선/기선…은 흘수에 의하여 제한을 받음.
.20 최대 허용 흘수는 …미터임.
.21 귀선은 횡경사가 있는가?
.21.1 예, 본선은 좌현/우현으로 …도 횡경사 되어 있음.
.21.2 아니오, 본선은 횡경사가 없음.
.22 귀선은 등흘수인가?
.22.1 예, 본선은 등흘수임.
.22.2 아니오, 본선은 선수 쪽/선미 쪽으로 트림이 있음.

.2 조난 통화 정보의 획득과 제공

AI/1.1 "조난 통신" 참조

AI/6.2 선박 통항 관제 업무(VTS)를 제공하기 위한 용어

.1 정보제공 서비스

이 어귀는 통상적으로 육지에서 발신됨.

.1.1 항행 경보

.1 위치 …에 정체불명의 물체가 있음.

.2 Ice/iceberg(s) in position …/area around … .
.3 Unlit derelict vessel adrift in vicinity … at …(*date and time*).
.4 Dangerous wreck/obstruction located in position … marked by …(*type*) buoy.
.5 Hazardous mine adrift in vicinity … at …(*date and time*).
.6 Uncharted reef/rock/shoal reported in position … .
.7 Pipeline is leaking gas/oil in position … - wide berth requested.
.8 Depth of water not sufficient in position … .
.9 Navigation closed in area … .

.1.2 Navigational information

.1 Oil spill in position … .
.2 Current meters/hydrographic instruments moored in position … - wide berth requested.
.3 Platform … (*name/number*) reported/established in position … - wide berth requested.
.4 …(*charted name of light/buoy*) in position …
~ unlit/unreliable/damaged/destroyed/off station/missing.
~ (temporarily) changed to …(*full characteristics*).
~ (temporarily) removed.
~ (temporarily) discontinued.
.5 …(*charted name of light/buoy*) …(*full characteristics*)
~ established in position … .
~ re-established in position … .
~ moved … kilometres/nautical miles in … (*direction*) to position … .
.6 (Note: Only for major fog signal stations.)
Fog signal …(*charted name of light/buoy*) in position … inoperative.

.2 위치 …에/…근처의 해역에 얼음/빙산이 있음.
.3 …근처, …(일시)에 소등된 유기선이 표류중임.
.4 위험한 난파선/장애물이 위치 …에 …(형식) 부표 부착되어 존재함.
.5 위험한 기뢰가 …근처 …(일시)에 표류하고 있음.
.6 위치 …에 해도에 나와 있지 않은 암초/바위/사주가 보고됨.
.7 위치 …에서 파이프라인이 가스/유류를 누출하고 있음 - 넓은 여유 수면이 요구됨.
.8 위치 …에 수심이 충분치 않음.
.9 해역 …의 항해는 금지되었음.

.1.2 항행 정보

.1 위치 …에 유출된 유류가 있음.
.2 위치 …에 해류계/수로 측량 기구가 계류되어 있음.
-넓은 여유수면이 요구됨.
.3 플랫폼 …(명칭/번호) 이 위치 …에 보고됨/설치됨 -넓은 여유수면이 요구됨.
.4 위치 …의 …(등대/부표의 해도상 명칭)이 소등됨/신뢰할 수 없음/손상됨/파괴됨/제 위치에서 이탈함/소실됨.
위치 …의 …(등대/부표의 해도상 명칭)이 …(상세 특성) 으로 (일시)변경되었음.
위치 …의 …(부표의 해도상 명칭)이 (일시)제거됨.
위치 …의 …(등대/부표의 해도상 명칭)이 (일시)중지됨.
.5 위치 …에 …(상세 특성)의 …(등대/부표의 해도상 명칭)이 설치되었음.
위치 …에 …(상세 특성)의 …(등대/부표의 해도상 명칭)이 재설치 되었음.
…(상세 특성)의 …(등대/부표의 해도상 명칭)이 …킬로미터/해리 …도 방향의 위치 …로 이동됨.
.6 (주: 주요 무중신호소에 국한함)
위치 …의 무중신호 …(등대/부표의 해도상 명칭)가 고장임

.1.3 Traffic information

.1 Gunnery/rocket firing/missile/torpedo/underwater ordnance exercises in area bounded by …(*positions*) from …(*date and time*) to …(*date and time*).
- Wide berth requested.

.2 Cable/pipeline operations by …(*vessel*) in vicinity … /along a line joining …(*positions*) from …(*date and time*) to …(*date and time*)
- wide berth requested.
Contact via VHF Channel …

.3 Salvage operations in position … from … (*date and time*) to … (*date and time*) - wide berth requested.[75)] Contact via VHF Channel … .

.4 Seismic/hydrographic operations by …(*vessel*) from … (*date and time*) to …(*date and time*) in position …
- wide berth requested. Contact via VHF Channel … .

.5 Oil clearance operations near MT … in position …
- wide berth requested.

.6 Transhipment of …(*kind of cargo*) in position …
- wide berth requested.

.7 Difficult tow from …(*port of departure*) to …(*destination*) on …(*date*) - wide berth requested.

.8 Vessel not under command in position …/area … .

.9 Hampered vessel in position …/area … (course … degrees, speed … knots).

.9.1 Vessel constrained by her draft in position …/area… (course… degrees, speed … knots)

.10 Vessel in position … on course … and at speed … is not complying with traffic regulations.

.11 Vessel is crossing … traffic lane on course … and speed … in position … .

.12 Small fishing boats in area around … - navigate with caution.

.1.3 통항 정보

.1 … , …위치(들)로 경계 지워진 해역에서 …(일시)부터 …(일시)까지 사격/로켓 발사/미사일/어뢰/해저 무기 훈련이 행하여짐. 넓은 여유수면이 요구됨.

.2 케이블/파이프라인 작업이 …(선명)에 의하여, …(일시) 부터…(일시)까지 …근처에서/…(위치)에서 만나는 선을 따라 이루어짐.
-넓은 여유수면이 요구됨.
VHF채널 …를 통하여 교신할 것.

.3 위치 …에서 …(일시)로부터 …(일시)까지 침몰선 인양 작업이 이루어짐.
-넓은 수역이 요구됨.
VHF채널 …를 통하여 교신할 것.

.4 …(선명)에 의하여, 지진 조사/수로 측량작업이 위치 …에서 …(일시)로부터 …(일시)까지 이루어짐.
-넓은 여유수면이 요구됨.
VHF 채널 …를 통하여 교신할 것

.5 위치 …의 탱커 …근처에서 유류 제거 작업이 진행됨.
-넓은 여유수면이 요구됨

.6 위치 …에서 …(화물의 종류)의 이적 작업이 이루어짐.
-넓은 여유수면이 요구됨

.7 …(일자)에, …(출항지)로부터 …(도착지)로 가는 힘든 예인 작업이 이루어짐. -넓은 여유수면이 요구됨

.8 위치 …/해역…에 운전 부자유선이 있음.

.9 위치 …/해역…에 조종 성능 제한선이 있음.(침로…도, 속력 …노트)

.9.1 위치 …/해역…에 흘수로 제한 받는 선박이 있음.(침로…도, 속력 …노트)

.10 위치 … , 침로 … , 속력 …의 선박은 통항 규칙을 위반하고 있음.

.11 위치 …에 침로…, 속력…의 …일방 통항로를 횡단하는 선박 있음.

.12 …주위의 해역에 소형 어선들이 있음. -주의해 항해하시오.

.13 Submarines operating in sea area around … , surface vessels are in attendance.

.1.4 Route information

.1 Route …/Traffic Lane … has been suspended/discontinued/diverted.

.1.5 Hydrographic information

.1 Tidal prediction for …(*name of station(s)*)/area … :

.1.1 A tide of … metres above/below datum is expected in position …/area … at about … UTC.

.1.2 Abnormally high/low tides are expected in position … /area … at about … UTC.

.2 The tide is rising -

~ it is … hours before high water/after low water.

~ it is … metres below high water/above low water.

.3 The tide is falling -

~ it is … hours after high water/before low water.

~ it is … metres below high water/above low water.

.4 The tide is slack.

.5 Present tide is … metres above/below datum in position … .

.6 The tide is … metres above/below prediction.

.7 The tidal stream/current is … knots in position … .

.8 The tide is setting in direction … degrees.[76)]

.9 The depth of water is/is not sufficient in position … .

.10 Charted depth has increased/decreased by … metres due to winds/sea state.

.1.6 Electronic navigational aids information

.1 GPS Satellite …(*number*) unusable from …(*date and time*) to .. (*date and time*).

.13 …주위의 해역에서 잠수함이 작전 중이며, 수상함들도 참가 중임.

.1.4 항로 지정 통보

.1 지정항로 …/일방 통항로 …는 일시 중지됨/중지됨/변경됨

.1.5 수로 통보

.1 …(국의 명칭)/…해역에 대한 조석 예보:

.1.1 위치…/해역…에서 세계시…시경 기본 수준면 상/하 …미터의 조석이 예상됨.

.1.2 위치 …/해역…에서 세계시 …시경에 이상 고조/저조 예상됨.

.2 조석은 창조(상승중)임 -고조시 이전/저조시 이후 …시간임
조석은 창조(상승중)임 -고조면하/저조면상…미터임.

.3 조석은 낙조(하강중)임 -고조시 …시간 후임/저조시 …시간 전임.
조석은 낙조(하강중)임 -고조면 하 …미터임/저조면상 …미터임.

.4 조석은 게조임.

.5 위치 …에서 현재의 조위는 기본수준면 상/하 …미터임.

.6 조위는 예보보다 …미터 높음/낮음.

.7 위치…에서 조류/해류는 …노트임.

.8 조류는…도 방향으로 흐르고 있음.

.9 위치…에서 수심은 충분함/충분치 않음.

.10 바람/해면 상태 때문에 해도에 기재된 수심은…미터 증가함/감소함.

.1.6 전자 항행 보조 설비 관련 통보

.1 GPS위성…(번호)은 …(일시)부터…(일시)까지 사용 불가능

Cancel one hour after time of restoration.

.2 LORAN station …(*name/number of master/secondary*) off air from … (*date and time*) to …(*date and time*).

.3 RACON …(*name of station*) in position … off air from …(*date and time*) to …(*date and time*).

.1.7 Meteorological warnings

.1 Gale warning/storm warning was issued at … UTC starting at … UTC.[77)]

.1.1 Gale warning/storm warning. Wind at … UTC in area …(*met. area*)from direction …(*cardinal points*) and force Beaufort … backing/veering to …(*cardinal points*).

.2 Tropical storm warning was issued at … UTC starting at … UTC.

.2.1 Tropical storm warning at … UTC. Hurricane …(*name*)/ tropical cyclone/tornado/willy-willy/typhoon/… with central pressure of …millibars/hPascals located in position … . Present movement …(*cardinal points*) at … knots. Winds of … knots within radius of … nautical miles of centre. Seas over … metres. Further information on VHF Channel …/frequency …(at …UTC).

.1.8 Meteorological information

.1 Position of tropical storm …(*name*) … , path …(*cardinal points*), speed of advance … knots.

.2 Wind direction …(*cardinal points*), force Beaufort … in position…

.3 Wind is backing/veering and increasing/decreasing.

.4 Wind is expected to increase/decrease in position … to force Beaufort … within the next … hours.

.5 Visibility in position …

~ … metres/nautical miles.

~ reduced by mist/fog/snow/dust/rain/… .

복구된 시각으로부터 1시간 후 취소함.

.2 LORAN국…(주국/종국의 명칭/번호)은 …(일시)부터…(일시) 까지 방송 중지 됨.

.3 위치…의 RACON…(국명)은… (일시)부터…(일시)까지 방송 중지됨.

.1.7 기상 경보

.1 세계시…시에 시작되는 강풍 경보/폭풍 경보가…(세계시)에 발령 되었음.

.1.1 강풍 경보/폭풍 경보. 세계시…시, …해역의 바람은 풍향… (방점), 풍력계급…으로부터 …(방점) 으로 반전/순전함.

.2 세계시…시에 시작되는 열대성 폭풍 경보가…(세계시)에 발령 되었음.

.2.1 세계시…에서의 열대성 폭풍 경보.

허리케인…(이름)/열대성사이클론/토네이도/윌리윌리/태풍/… 의 중심 기압은…밀리바/헥토파스칼이고 …에 위치함.

현재의 이동 방향은…(방점), …노트임. 중심으로부터 반경… 해리 이내의 풍속은… 노트임. 파고는 … 미터 이상임.

추가통보가 VHF 채널…/주파수…에서(세계시…시에) 있음.

.1.8 기상 통보

.1 열대성 폭풍…(명칭)의 위치는 … , 이동 경로는…(방점), 전진 속도는…노트임.

.2 위치…에서 풍향은…(방점) 풍력계급은…임.

.3 풍향은 반전/순전하며 강해지고/약해지고 있음.

.4 위치…에서의 바람은 앞으로 …시간 이내에 풍력계급…으로 증가/감소하겠음.

.5 위치…에서의 시정은…미터/해리임.

위치…에서의 시정은 엷은 안개/안개/눈/먼지/강우/…로 인하여 감소하였음.

~ expected to increase/decrease to … metres/ nautical miles within the next … hours.

.6 Sea/swell in position …

~ … metres from …(*cardinal points*).

~ expected to increase/decrease within the next … hours.

.7 Icing is expected/not expected in area … .

.1.9 Meteorological questions and answers

See AI/3.1 "Meteorological and hydrological conditions"

.2 Navigational assistance service

Shore based pilotage by Navigational Assistance Service: also see AI/6.4 .3.18 to .3.21

.2.1 Request and identification

.1 Is shore based radar assistance available?

.1.1 Yes, shore based radar assistance is available.

.1.2 No, shore based radar assistance is not available.

.2 Shore based radar assistance is available from … to … UTC.

.3 Do you require navigational assistance to reach … ?

.3.1 Yes, I require navigational assistance.

.3.2 No, I do not require navigational assistance.

.4 What is your position?

.4.1 My position is bearing … degrees, distance … kilometres/ nautical miles from … .

.5 How was your position obtained?

.5.1 My position was obtained by GPS/RADAR/cross-bearing/ astronomical observation/… .

.6 Repeat your position for identification.

.7 I have located you on my radar screen.

위치…에서의 시정은 앞으로 …시간 이내에…미터/해리로 증가/감소하겠음.

.6 위치…에서의 파도/너울은 높이…미터이고 …(방점)에서 오고 있음. 위치…에서의 파도/너울은 앞으로 …시간 이내에 더 세어/약해질 것임.

.7 …해역에 결빙이 예상됨/예상되지 않음.

.1.9 기상 관련 문답

AI/3.1 "기상 및 해상 상태“ 참조

.2 항행 원조 업무

항행 원조 업무로서의 육상 기지의 도선 업무에 관하여는 AI/6.4. 3.18에서 3.21를 참조

.2.1 요청 및 식별

.1 해안 레이더의 원조는 이용 가능한가?

.1.1 예, 해안 레이더의 원조는 이용 가능함.

.1.2 아니오, 해안 레이더의 원조는 이용 가능하지 않음.

.2 세계시 …시에서 …시까지 해안 레이더 원조 이용 가능함.

.3 귀선은 …에 도착하기 위하여 항행 원조를 필요로 하는가?

.3.1 예, 본선은 항행 원조를 필요로 함.

.3.2 아니오, 본선은 항행 원조를 필요로 하지 않음.

.4 귀선의 위치는 어디인가?

.4.1 본선의 위치는 …로부터 방위 …도 방향, 거리 …킬로미터/ 해리임.

.5 귀선의 위치는 어떻게 구하였는가?

.5.1 본선의 위치는 GPS/RADAR/교차 방위법/천측/…으로 구했음

.6 식별하기 위하여 귀선의 위치를 반복하시오.

.7 이쪽 레이더 화면상에서 귀선을 위치 확인했음.

.7.1 Your position is bearing … degrees, distance … kilometres/ nautical miles from … .
.8 I cannot locate you on my radar screen.
.9 What is your present course and speed?
.9.1 My present course is … degrees, my speed is … knots.
.10 What is the course to reach you?
.10.1 The course to reach me is … degrees.
.11 Is your radar in operation?
.11.1 Yes, my radar is in operation.
.11.2 No, my radar is not in operation.
.12 What range scale are you using?
.12.1 I am using … miles range scale.
.12.2 Change to a larger/smaller range scale.
.13 You are leaving my radar screen.
.14 Change to radar …(*name*) , VHF Channel … .
.15 I have lost radar contact.

.2.2 Position

.1 You are entering … .
.2 Your position is bearing … degrees, distance … kilometres/nautical miles from … .
.4 You are passing … .
You are
~ in the centre of the fairway.
~ on/not on the radar reference line (of the fairway).
~ on the …(*cardinal points*) side of the fairway.
.5 You are approaching the …(*cardinal points*) limit of the fairway.
.6 Your position is buoy number … distance … metres/ cables to the …(*cardinal points*) of the radar reference line.
.7 Your position is distance … metres/cables from the intersection

.7.1 귀선의 위치는 …로부터 방위 …도, 거리 …킬로미터/해리임.
.8 귀선을 이쪽 레이더 화면상에서 위치 확인할 수 없음.
.9 귀선의 현재 침로와 속력은 얼마인가?
.9.1 본선의 현재 침로는 …도, 속력은 …노트임.
.10 그 쪽에 도착하기 위한 침로는 몇 도인가?
.10.1 이 쪽에 도착하기 위한 침로는 …도임.
.11 귀선의 레이더는 작동중인가?
.11.1 예, 본선의 레이더는 작동중임.
.11.2 아니오, 본선의 레이더는 작동하지 않고 있음.
.12 귀선은 어떤 레인지 스케일을 사용 중인가?
.12.1 본선은 …마일 레인지 스케일을 사용 중임.
.12.2 더 큰/작은 레인지 스케일로 바꾸시오.
.13 귀선은 이쪽의 레이더 화면상에서 벗어나고 있음.
.14 레이더 …(명칭), VHF채널 …로 바꾸시오.
.15 이 쪽은 레이더에 의한 포착을 놓쳤음.

.2.2 위치

.1 귀선은 …에 진입하고 있음.
.2 귀선의 위치는 …로부터 방위…도, 거리…킬로미터/해리임.
.4 귀선은 …를 통과 중임.
귀선은 항로의 중앙에 있음.
귀선은 (항로의) 레이더 참조선상에 있음/있지 않음.
귀선은 항로의 …(방점)측에 있음.
.5 귀선은 항로의…(방점)측 한계에 접근하고 있음.
.6 귀선의 위치는 부표 번호 …번, 레이더 참조선의 …(방점)측 까지 거리 …미터/케이블임.
.7 귀선의 위치는 레이더 참조선…과 레이더 참조선…의 교차점으로부터 거리…미터/케이블이고,

of radar reference line … and radar reference line … and distance … metres/cables to the …(*cardinal points*) of radar reference line … .

.8 MV … has reported at reporting point … .

.9 You are getting closer to the vessel …(*cardinal points*) of you.

.10 Vessel on opposite course is passing to the …(*cardinal points*) of you.

.11 MV … , … metres/cables …(*cardinal points*) of you
- ~ is ingoing/outgoing.
- ~ has stopped.
- ~ is at anchor.
- ~ is on a reciprocal course.
- ~ will overtake to the …(*cardinal points*) of you.

.12 Vessel has anchored … metres/cables …(*cardinal points*) of you in position … .

.13 Vessel …(*cardinal points*) of you is obstructing your movements.

.14 You will meet crossing traffic in position … .

.15 Vessel is entering/leaving the fairway at … .

.16 Buoy … is distance … metres/cables … (*cardinal points*).

.17 Vessel … (*cardinal points*) of you is
- ~ turning.
- ~ anchoring.
- ~ increasing/decreasing speed.
- ~ overtaking you.
- ~ not under command.

.2.3 Course

Note: The user of this phrase should be fully aware of the implications of words such as "track", "heading" and "course made good".

.1 Your track is

레이더 참조선…의 …(방점)측까지 거리가 …미터/케이블임.

.8 기선…이 보고 지점…에서 보고하였음.

.9 귀선은 귀선의 …(방점)쪽의 선박에 접근하고 있음.

.10 반대 침로의 선박은 귀선의 …(방점)측을 통과중임.

.11 귀선의…(방점)쪽, …미터/케이블에 있는 기선…은 입항/출항 중임.
귀선의…(방점)쪽, …미터/케이블에 있는 기선…은 정지중임.
귀선의…(방점)쪽, …미터/케이블에 있는 기선…은 투묘중임.
귀선의…(방점)쪽, …미터/케이블에 있는 기선…은 반대 침로임.
귀선의…(방점)쪽, …미터/케이블에 있는 기선…은 귀선의… (방점)측을 추월할 것임.

.12 귀선의…(방점)측…미터/케이블의 위치…에 선박이 투묘중임.

.13 귀선의 …(방점)측에 있는 선박이 귀선의 이동을 방해하고 있음.

.14 귀선은 위치…에서 횡단선을 만나게 될 것임.

.15 …에서 선박이 항로에 진입중임/떠나고 있음.

.16 부표…이 거리…미터/케이블, …측(방점)에 있음.

.17 귀선의 …(방점)측에 있는 선박이 선회중임
귀선의 …(방점)측에 있는 선박이 투묘중임
귀선의 …(방점)측에 있는 선박이 증속중/감속중임.
귀선의 …(방점)측에 있는 선박이 귀선을 추월중임
귀선의 …(방점)측에 있는 선박이 운전이 부자유함.

.2.3 침로

주: 이 어귀를 사용하는 사람은 “track(진로)” “heading(선수방위)” “course made good"(대지침로)과 같은 단어의 의미를 완전히 알고 있어야 한다.

.1 귀선의 진로는 참조선에 평행임.

~ parallel with the reference line.
~ diverging from the reference line.
~ converging to the reference line.

.2 What is your present course/heading?

.2.1 My present course/heading is … degrees

.3 You are steering a dangerous course.

.4 Course to make good is … degrees.

.5 Vessel …*(cardinal points)* of you is on same course … degrees.

.5.1 Advise you
~ keep your present course.
~ a new course of … degrees.

.6 Have you altered course?

.6.1 Yes, I have altered course - my new course is … degrees.

.6.2 No, I have not altered course - my course is … degrees.

.7 You are running into danger -
~ shallow water …*(cardinal points)* of you.
~ submerged wreck …*(cardinal points)* of you.
~ fog bank …*(cardinal points)* of you.
~ risk of collision (with a vessel bearing … degrees, distance … kilometres/nautical miles).
~ bridge is defective/… .

.3 Traffic organization service

.3.1 Clearance, forward planning

.1 Traffic clearance is required before entering … .

.2 Do not enter the traffic lane/… .

.3 Proceed to the emergency anchorage.

귀선의 진로는 참조선으로부터 벗어나고 있음.
귀선의 진로는 참조선에 접근하고 있음.

.2 귀선의 현재 침로/선수 방위는 몇 도인가?

.2.1 본선의 현재 침로/선수 방위는…도임.

.3 귀선은 위험 침로로 조타하고 있음.

.4 올바른 침로는 …도임.

.5 귀선의 …(방점)측에 있는 선박은 동일 침로…도임.

.5.1 귀선의 현재 침로를 유지하기를 권고함.
신 침로 …도를 권고함.

.6 귀선은 변침하였는가?

.6.1 예, 본선은 변침하였음 - 본선의 신 침로는 …도임.

.6.2 아니오, 본선은 변침하지 않았음 - 본선의 침로는 …도임.

.7 귀선은 위험에 빠져들고 있음- 귀선의…(방점)측에 수심이 얕은 해역이 있음.
귀선은 위험에 빠져들고 있음- 귀선의 …(방점)측에 물 속에 가라앉은 난파선이 있음.
귀선은 위험에 빠져들고 있음- 귀선의…(방점)측에 무제 (농무해역)가 있음.
귀선은 위험에 빠져들고 있음- (방위…도, 거리… 킬로미터/ 해리의 선박과)충돌의 위험이 있음.
귀선은 위험에 빠져들고 있음- 교량이 고장/…임.

.3 통항 편성 업무

.3.1 허가, 진행 계획

.1 …에 진입하기 전에 통항 허가가 요구됨.

.2 일방 통항로/…에 진입하지 마시오.

.3 비상 투묘지로 항진하시오.

.4 Keep clear of …/ avoid … .
.5 You have permission
 ~ to enter the traffic lane/route - traffic clearance granted.
 ~ to enter traffic lane/route in position … at … UTC.
.6 Do not pass the reporting point … until … UTC.
.7 Report at the next way point/way point …/at … UTC.
.8 You must arrive at way point … at … UTC - your berth is clear.
.9 Do not arrive in position … before/after … UTC.
.10 The tide is with you/against you.

.3.2 Anchoring

.1 You must anchor
 ~ at … UTC.
 ~ until the pilot arrives.
 ~ in a different position.
 ~ clear of fairway.
.2 Do not anchor in position … .
.3 Anchoring is prohibited.
.4 You must heave up anchor.
.5 You are at anchor in a wrong position.
.6 Have your crew on stand by for heaving up anchor when the pilot embarks.
.7 You have permission to anchor (at … UTC)
 ~ in position … .
 ~ until the pilot arrives.
 ~ until the tugs arrive.
 ~ until sufficient water.
.8 You are obstructing the fairway/other traffic.
.9 Are you dragging/dredging anchor?

.4 …에 접근하지 마시오/…을 피하시오.

.5 귀선은 일방 통항로/지정항로에 진입하여도 됨-통항이 허가되었음.
귀선은 세계시…시에 위치…에서 일방 통항로/지정항로에 진입해도 됨.

.6 세계시…시까지는 보고 지점…를 통과하지 마시오

.7 다음의 통과 지점/통과 지점…에서/세계시…시에 보고하시오

.8 귀선은 세계시…시에 통과 지점…에 도착해야함- 귀선의 정 박지는 비어있음

.9 세계시…시 이전/이후에는 위치…에 도착하지 마시오

.10 조류는 순조임/역조임

.3.2 투묘

.1 귀선은 세계시 …시에 투묘해야 함.
귀선은 도선사가 도착할 때까지 투묘해야 함.
귀선은 다른 위치에 투묘해야 함.
귀선은 항로에서 떨어져서 투묘해야 함.

.2 위치 …에 투묘하지 마시오.

.3 투묘는 금지되어 있음.

.4 귀선은 양묘하여야 함.

.5 귀선은 틀린 장소에 묘박중임.

.6 도선사가 승선하면 양묘할 수 있도록 귀선의 선원들을 대기 시키시오.

.7 귀선은 위치 …에 (세계시 …시에)투묘가 허락됨.
귀선은 도선사가 도착할 때까지 (세계시 …시에)투묘 허락됨.
귀선은 예인선이 도착할 때까지 (세계시 …시에)투묘 허락됨.
귀선은 수심이 충분할 때까지 (세계시 …시에)투묘 허락됨.

.8 귀선은 항로/타선의 통항을 가로막고 있음.

.9 귀선은 주묘중/닻을 끌고 있는 중인가?

.9.1 Yes, I am dragging/dredging anchor.

.9.2 No, I am not dragging/dredging anchor.

.10 Do not dredge anchor.

.3.3 Arrival, berthing and departure

.1 Your orders are to berth on … .

.2 Your orders are changed to proceed to … .

.3 Proceed to … for orders.

.4 You have permission to enter/to proceed at … UTC.

.5 Vessel is turning/manoeuvring in position … .

.6 MV …

~ will turn in position … .

~ will leave … at … UTC.

~ is leaving … .

~ has left … .

~ entered fairway in position … .

.7 Your berth is not clear (until … UTC).

.7.1 Your berth will be clear at … UTC.

.8 You will berth/dock at … UTC .

.9 Berthing has been delayed by … hours.

.10 Be ready to get underway.

.10.1 I am ready to get underway.

.11 Get underway.

.12 Are you underway?

.12.1 Yes, I am underway.

.12.2 No, I am not underway.

.13 Move ahead/astern … metres.

.14 Your vessel is in position - make fast.

.9.1 예, 본선은 주묘중/닻을 끌고 있는 중임.
.9.2 아니오, 본선은 주묘중/닻을 끌고 있는 중이 아님.
.10 닻을 끌지 마시오.

.3.3 입항, 정박 및 출항

.1 귀선에 대한 지시는 …에 정박하라는 것임.
.2 귀선에 대한 지시는 …로 항진하라고 변경되었음.
.3 …에 항진하여 지시를 받으시오.
.4 귀선은 세계시 …시에 진입/항진하도록 허락됨.
.5 위치 …에서 선박이 선회중/조종중임.
.6 기선… 이 위치 …에서 선회할 것임.
기선… 이 세계시 …시에 …를 떠날 것임.
기선… 이 …를 떠나고 있음.
기선… 이 …를 떠났음.
기선…이 위치 …에서 항로에 진입하였음.
.7 (세계시 …시까지)귀선의 정박지는 비어있지 않음.
.7.1 귀선의 정박지는 세계시 …시에 비게 될 것임.
.8 귀선은 세계시…시에 정박/입거할 것임.
.9 정박은 …시간 연기되었음.
.10 항해를 개시할 준비를 하시오.
.10.1 본선은 항해를 개시할 준비가 되었음.
.11 항해를 개시하시오.
.12 귀선은 항해중인가?
.12.1 예, 본선은 항해중임.
.12.2 아니오, 본선은 항해중이 아님.
.13 전/후방으로 …미터 이동하시오.
.14 귀선은 정 위치임- 잡아매시오.

.3.4 Enforcement

.1 According to my radar, your course does not comply with Rule 10 of the COLREGs.[78)]

.2 Your actions will be reported to the Authorities.

.3 You are

~ not complying with traffic regulations.

~ not keeping to the correct traffic lane.

.4 Have all navigational instruments in operation before entering this area/area … .

.5 Your navigation lights are not visible.

.6 Recover your fishing gear.

.6.1 You are fishing in the fairway.

.7 Fishing gear is to the …(*cardinal points*) of you.

.8 Fishing in area … is prohibited.

.9 You are approaching a prohibited fishing area.

.10 Fairway speed is… knots.

.3.5 Avoiding dangerous situations, providing safe movements

.1 It is dangerous

~ to anchor in your present position.

~ to remain in your present position.

~ to alter course to …(*cardinal points*).

.2 Large vessel is leaving the fairway - keep clear of the fairway approach.

.3 Nets with buoys/without buoys in this area - navigate with caution.

.4 Collision in position … .

.5 MV … is aground/on fire/… in position … .

.6 Stand by for assistance.

.7 Vessels must

~ keep clear of this area/area … .

.3.4 집 행

.1 이쪽의 레이더에 의하면, 귀선의 침로는 국제 해상 충돌 예방규칙의 제 10조를 따르지 않고 있음.

.2 귀선의 행위는 당국에 통보될 것임.

.3 귀선은 통항 규칙을 지키지 않고 있음.
귀선은 올바른 일방 통항로를 따라가지 않고 있음.

.4 이 해역/…해역에 진입하기 전에 모든 항해 장비를 작동 시키시오.

.5 귀선의 항해등은 보이지 않음.

.6 귀선의 어구를 걷어 들이시오,

.6.1 귀선은 항로 내에서 어로 작업 중임.

.7 어구는 귀선의 …(방전)측에 있음.

.8 …해역에서의 어로는 금지되었음.

.9 귀선은 어로 금지 구역에 접근 중임.

.10 항로 내의 제한속력은 …노트임.

.3.5 위험한 상황의 회피, 안전한 통항의 제공

.1 귀선의 현재 위치에 투묘하는 것은 위험함.
귀선의 현재 위치에 머무르는 것은 위험함.
…(방점)측으로 변침하는 것은 위험함.

.2 대형선이 항로에서 벗어나고 있음
- 항로 접근로에서 떨어지시오.

.3 이 해역에 부표 달린/부표 없는 어망이 있음
- 주의 항해하시오.

.4 위치 …에서 충돌 사고 발생.

.5 기선…이 위치 …에서 좌초/화재 발생/…하였음.

.6 원조를 준비를 하시오.

.7 선박들은 이 해역/…해역에 접근하지 말 것.

~ avoid this area/area … .

~ navigate with caution.

.8 Keep clear of … - search and rescue in progress.

.9 Your present course is too close

~ to ingoing/outgoing vessel.

~ to the vessel that you are overtaking.

~ to the …(*cardinal points*) limit of the fairway.

.10 Your course is deviating from the radar reference line.

.11 You are running into danger

~ shallow water …(*cardinal points*) of you.

~ submerged wreck …(*cardinal points*) of you.

~ fog bank …(*cardinal points*) of you.

~ risk of collision (with vessel bearing … degrees, distance … kilometres/nautical miles).

~ bridge is defective.

.12 You are proceeding at a dangerous speed.

.13 You must

~ proceed by the fairway/route … .

~ keep to the …(*cardinal points*) of the fairway line/ radar reference line.

~ stay clear of the fairway.

.14 You must wait for MV … to cross ahead of you.

.15 You must wait for MV … to clear … before

~ entering the fairway.

~ getting underway.

~ leaving the berth.

.16 Do not

~ overtake.

~ cross the fairway.

.17 Alter course to …*(cardinal points)* of you.

.18 Pass …*(cardinal points)* of

선박들은 이 해역/…해역에서 피해야 함.
선박들은 주의하여 항해하여야 함.

.8 …에 접근하지 마시오 - 수색과 구조 작업이 진행 중임.

.9 귀선의 현재 침로는 입항선/출항선에 너무 접근함.
귀선의 현재 침로는 귀선이 추월하고 있는 선박에 너무 접근함.
귀선의 현재 침로는 항로의 …(방점)측 한계선에 너무 접근함.

.10 귀선의 침로는 레이더 참조선으로부터 벗어나고 있음.

.11 귀선은 위험에 빠져들고 있음, 귀선의…(방점)측에 수심이 얕은 해역이 있음.
귀선은 위험에 빠져들고 있음, 귀선의 …(방점)측에 물속에 가라앉은 난파선이 있음.
귀선은 위험에 빠져들고 있음, 귀선의…(방점)측에 농무해역 있음
귀선은 위험에 빠져들고 있음, (방위…도, 거리… 킬로미터/해리의 선박과)충돌의 위험이 있음.
귀선은 위험에 빠져들고 있음, 교량에 결함이 있음.

.12 귀선은 위험 속력으로 항진중임.

.13 귀선은 …항로/지정항로로 항진하여야함.
귀선은 항로선/레이더 참조선의…(방점)쪽을 유지하여야 함.
귀선은 항로에서 떨어져 있어야 함.

.14 귀선은 기선…이 귀선의 전방을 횡단하도록 기다려야 함.

.15 귀선은 항로에 진입하기 전에 기선…이 …를 벗어날 때까지 기다리시오
귀선은 항해를 개시하기 전에 기선…이 …를 벗어날 때까지 기다리시오.
귀선은 정박지를 떠나기 전에 기선…이 …를 벗어날 때까지 기다리시오.

.16 추월하지 마시오.
항로를 횡단하지 마시오.

.17 귀선의 …(방점)측으로 변침하시오.

.18 입항 중/출항 중/투묘 중/파손된 선박의

~ ingoing /outgoing/anchored/disabled vessel.
~ ··· mark/··· .

.19 Stop engines.

.20 MV ···
~ wishes to overtake ···*(cardinal points)* of you.
~ agrees/does not agree to be overtaken.
~ is approaching an obscured area ···
- approaching vessels acknowledge.

.3.6 Canal and lock operations

.1 You must
~ close up on the vessel ahead of you.
~ drop back from the vessel ahead of you.
~ wait at ··· .
~ moor at ··· .
~ wait for lock clearance at ··· until ··· UTC.

.2 Convoy ··· must wait/moor at ··· .

.3 You will
~ join convoy ··· at ··· UTC.
~ enter canal/lock at ··· UTC.

.4 Transit will begin at ··· UTC.

.5 Your place in convoy is number ··· .

.6 Transit/convoy speed is ··· knots.

.7 Convoys/vessels will pass in area ··· .

AI/6.3 Handing over to another VTS

.1 ··· VTS this is ··· VTS: MV ··· position is bearing ··· degrees, distance ···kilometres/nautical miles from ··· . Working frequency is VHF Channel ··· . Your target. Please confirm.

.2 ··· VTS this is ··· VTS: MV ··· position is bearing ··· degrees, distance ···kilometres/nautical miles from ··· . I confirm. My target.

…(방점)쪽을 통과 하시오.
…표지/…의 …(방점)쪽을 통과하시오.
.19 기관을 정지하시오.
.20 기선…은 귀선의 …(방점)측을 통하여 추월하기를 원함.
기선…은 추월 당하는데 동의함/동의하지 아니함.
기선…이 불명료 구역…에 접근하고 있음
- 접근중인 선박들은 주지하기 바람.

.3.6 운하 및 갑문 운항

.1 귀선은 귀선 전방의 선박에 접근하여야 함.
귀선은 귀선 전방의 선박으로부터 멀어져야 함.
귀선은 …에서 기다려야 함.
귀선은 …에 계류하여야 함.
귀선은 …에서 세계시…시까지 갑문이 비기를 기다려야 함.
.2 호송선단 …은 …에서 기다려야/계류해야 함.
.3 귀선은 세계시…시에 호송 선단 …에 합류할 것임.
귀선은 세계시…시에 운하/갑문에 진입할 것임.
.4 통과가 세계시 …시에 개시될 것임.
.5 호송 선단 내에서 귀선의 위치는 …번째임.
.6 통과/호송 선단 속력은 …노트임.
.7 호송 선단/선박들은 …해역을 통과할 것임.

AI/6.3 다른 VTS 에로의 인계

.1 …VTS, 여기는 …VTS: 기선…의 위치는 …로부터 방위 …도, 거리 …해리이고, 교신주파수는 VHF 채널…임. 귀국의 물표임. 확인하기 바람.
.2 …VTS, 여기는 …VTS: 기선…의 위치는 …로부터 방위 … 도, 거리 …킬로미터/해리임. 본국은 확인함. 본국의 물표임.

.3 … VTS this is … VTS: MV … position is bearing… degrees, distance …kilometres/nautical miles from … . I am unable to take over this target.

AI/6.4 Phrases for communication with emergency services and allied services

.1 **Emergency services** (SAR, fire fighting, pollution fighting)
See AI/1 "Distress Communication"

.2 **Tug services**
Also see AII/3.6 "Tug assistance"
.1 How many tugs do you require?
.1.1 I require … tug(s).
.2 You must take
~ … tug(s) according to port regulations.
~ … tug(s) fore and … tug(s) aft.
.3 Wait for the tug(s) in position … .
.4 The tugs will meet you in position … at … UTC.
.5 Tug services have been suspended until …*(date and time)*/ resumed on …*(date and time)*.

.3 **Pilot request**
.1 Must I take a pilot?
.1.1 Yes, you must take a pilot - pilotage is compulsory.[79)]
.1.2 No, you need not take a pilot.
.2 Do you require a pilot?
.2.1 Yes, I require a pilot.
.2.2 No, I do not require a pilot - I am holder of Pilotage Exemption Certificate (No. …).

.3 …VTS, 여기는 …VTS: 기선…의 위치는 …로부터 방위 …도, 거리 …킬로미터/해리임.
본국은 이 물표를 인수할 수가 없음.

AI/6.4 비상 업무 및 협조 업무 관련 통신 용어

.1 비상 업무 (수색 구조, 소화, 오염 방제)
AI/1의 "조난 통신"을 참고할 것

.2 예인선 업무
AII/3.6의 "예인선의 원조"도 참고할 것
.1 귀선은 몇 척의 예인선을 필요로 하는가?
.1.1 본선은 …척의 예인선을 필요로함.
.2 귀선은 항구 규칙에 의하여 …척의 예인선을 사용하여야 함.
귀선은 선수에 예인선 …척, 선미에 예인선 …척을 사용하여야함.
.3 위치…에서 예인선을 기다리시오.
.4 예인선은 세계시 …시에 위치…에서 귀선을 만날 것임.
.5 예인선의 업무는 …(일시)까지 일시 중지됨/ …(일시)에 재개함.

.3 도선 요청
.1 본선은 도선사를 승선시켜야 하는가?
.1.1 예, 귀선은 도선사를 승선시켜야 함 - 도선은 강제임.
.1.2 아니오, 귀선은 도선사를 승선시킬 필요가 없음.
.2 귀선은 도선사를 요청하는가?
.2.1 예, 본선은 도선사를 요청함.
.2.2 아니오, 본선은 도선사를 요청하지 않음- 본선은 도선 면제증서 (번호…)를 소지하고 있음.

.3 You are exempted from pilotage.
.4 Do you require a pilot at …*(name)* Pilot Station?
.4.1 Yes, I require a pilot at …*(name)* Pilot Station.
.4.2 No, I do not require a pilot at …*(name)* Pilot Station - I require a pilot in position … .
.5 What is your ETA at …*(name)* Pilot Station in local time?
.5.1 My ETA at…*(name)* Pilot Station is … hours local time.
.6 What is local time?
.6.1 Local time is … hours.
.7 What is your position?
.7.1 My position is … .
.8 What is your distance from …*(name)* Pilot Station?
.8.1 My distance from …*(name)* Pilot Station is … kilometres/ nautical miles.
.9 Is the pilot boat on station?
.9.1 Yes, the pilot boat is on station.
.9.2 No, the pilot boat is not on station.
.9.3 The pilot boat will be on station at … hours local time.
.10 In what position can I take the pilot?
.10.1 Take the pilot at …*(Pilot Station)*/near … at … hours local time.
.11 When will the pilot embark?
.11.1 The pilot will embark at … hours local time.
.12 The pilot boat is coming to you.
.13 Stop in present position and wait for the pilot.
.14 Keep the pilot boat …*(cardinal points)* of you.
.15 What is your freeboard?
.15.1 My freeboard is … metres.
.16 Change to VHF Channel … for pilot transfer.
.17 Stand by on VHF Channel … until pilot transfer is completed.
.18 Pilotage at …*(name)* Pilot Station has been suspended until … *(date and local time)*.

.3 귀선은 도선 면제되었음.
.4 귀선은…(명칭) 파일럿 스테이션에서 도선사를 요청하는가?
.4.1 예, 본선은 …(명칭) 파일럿 스테이션에서 도선사를 요청함.
.4.2 아니오, 본선은 …(명칭) 파일럿 스테이션에서 도선사를 요청 하지 않음 - 본선은 위치 …에서 도선사를 요청함.
.5 귀선의 …(명칭)파일럿 스테이션에의 도착 예정 시각은 지방시로 몇 시인가?
.5.1 본선의 …(명칭)파일럿 스테이션 도착 예정 시각은 지방시로 …시임.
.6 지방시는 몇 시인가?
.6.1 지방시는 …시임.
.7 귀선의 위치는 어디인가?
.7.1 본선의 위치는 …임.
.8 …(명칭)파일럿 스테이션으로부터 귀선까시의 서리는 얼마인가?
.8.1 …(명칭)파일럿 스테이션으로부터 본선까지의 거리는 …킬로미터/해리 임.
.9 도선선은 정위치에 배치되어 있는가?
.9.1 예, 도선선은 정위치에 배치되어 있음.
.9.2 아니오, 도선선은 정위치에 배치되어 있지 않음.
.9.3 도선선은 지방시로 …시에 정위치에 배치되어 있게 됨.
.10 본선은 어디에서 도선사를 승선시킬 수 있는가.
.10.1 …(파일럿 스테이션)/…근처에서 지방시…시에 도선사를 승선시키시오.
.11 도선사는 몇 시에 승선하는가?
.11.1 도선사는 지방시…시에 승선할 것임.
.12 도선선은 귀선을 향해서 가고 있음.
.13 현재의 위치에 정지하여 도선사를 기다리시오.
.14 도선선을 귀선의…(방점)측에 위치시키시오.
.15 귀선의 건현은 얼마인가?
.15.1 본선의 건현은 …미터임.
.16 도선사의 이동을 위하여 VHF채널…으로 변경하시오.
.17 도선사의 이동이 완료될 때까지 VHF채널…에서 대기하시오.
.18 …(명칭)파일럿 스테이션에서의 도선 업무는…(날짜와 지방시)까지 일시 중지함.

.19 Pilotage at ···*(name)* Pilot Station has been resumed.

.20 The pilot cannot embark at ···*(name)* Pilot Station due to ··· .

.21 Do you accept shore-based navigational assistance from VTS Centre?

.21.1 Yes, I accept shore-based navigational assistance.

.21.2 No, I do not accept shore-based navigational assistance.

.21.3 I will stay in position ··· until ··· .

.22 You have permission to proceed by yourself (or wait for the pilot at ··· buoy).

.23 Follow the pilot boat inward where the pilot will embark.

.4 Embarking/disembarking pilot

See AI/4.2 "Embarking/disembarking pilot" (p.120)

.19 …(명칭)파일럿 스테이션에서의 도선 업무는 재개됨.

.20 …(명칭)파일럿 스테이션에서는 …로 인하여 도선사가 승선할 수 없음.

.21 귀선은 VTS 센터로부터의 육상 항행 원조를 받아들이는가?

.21.1 예, 본선은 육상 항행 원조를 받아들임.

.21.2 아니오, 본선은 육상 항행 원조를 받아들이지 않음.

.21.3 본선은 위치…에서 …까지 머무르겠음.

.22 귀선은 (도선사 없이) 자력으로 항진하여도 되고 (…부표에서 도선사를 기다려도 됨)

.23 안쪽으로 도선선을 뒤따르시오 그러면 거기서 도선사가 승선할 것임.

.4 도선사의 승선/하선

AI/4.2 "도선사의 승선/하선" 참조 **(p.120)**

Appendix to AI - External Communication Phrases

Standard GMDSS Messages

For further details see: ITU MANUAL for use by the Maritime Mobile and Maritime Mobile-Satellite Services, Geneva.[80)]

1 Standard distress message

.1 Structure

Upon receipt of a DSC Distress Alert acknowledgement the vessel in distress should commence the distress traffic on one of the international distress traffic frequencies such as VHF Channel 16 or frequency 2182 kHz (if not automatically controlled) as follows:[81)]

MAYDAY
THIS IS

- ☞ the 9-digit Maritime Mobile Service Identity number (MMSI) plus name/call sign or other identification of the vessel calling
- ☞ the position of the vessel
- ☞ the nature of distress
- ☞ the assistance required
- ☞ any other information which might facilitate rescue.

.2 Example

MAYDAY

- ☞ THIS IS TWO-ONE-ONE-TWO-THREE-NINE-SIX-EIGHT -ZERO MOTOR VESSEL "BIRTE" CALL SIGN DELTA ALPHA MIKE KILO
- ☞ POSITION SIX TWO DEGREES ONE ONE DECIMAL EIGHT MINUTES NORTH, ZERO ZERO SEVEN DEGREES FOUR FOUR MINUTES EAST

AI - 외부 통신 용어에 대한 부록

표준 GMDSS 통신문

더 자세한 사항은 : "해상 이동 통신 및 해상 이동 - 위성 통신 업무에 대한 ITU 지침, 제네바"를 참조할 것

1 표준 조난 통신문

.1 구성

DSC 조난경보 수신 확인 통부를 수신하면, 조난선우 VHF 채널 16번이나 2182kHz(자동으로 제어되지 않으면)와 같은 국제 조난 통화 주파수중 하나로 다음과 같이 조난 통화를 개시해야 한다:

<u>**메이 데이**</u>

여기는

- 9개의 숫자로 된 해상 이동통신업무 식별번호에 추가하여 선명/호출부호 또는 기타 호출 선박의 식별 부호
- 선박의 위치
- 조난의 성격
- 필요로 하는 원조
- 구조를 용이하게 할 수 있는 기타의 정보

.2 보기

<u>**메이데이**</u>

- 여기는 2-1-1-2-3-9-6-8-0 기선 "BIRTE"
  호출부호 Delta - Alpha - Mike - Kilo
- 위치는 북위 62도 11.8분, 동경 007도 44분

☞ I AM ON FIRE AFTER EXPLOSION
☞ I REQUIRE FIRE FIGHTING ASSISTANCE
☞ SMOKE NOT TOXIC OVER

2 Standard urgency message

.1 Structure

After the transmission of a DSC Urgency Call switch the transmitter to VHF Channel 16 or frequency 2182 kHz (if not automatically controlled) and commence the urgency traffic as follows:

PAN-PAN (repeated three times)
ALL STATIONS (repeated three times)
THIS IS

☞ the 9-digit MMSI of the vessel plus name/call sign or other identification
☞ the position of the vessel
☞ the text of the urgency message.

.2 Example

PAN-PAN PAN-PAN PAN-PAN
ALL STATIONS ALL STATIONS ALL STATIONS

☞ THIS IS
TWO-ONE-ONE-TWO-THREE-NINE-SIX-EIGH -ZERO MOTOR VESSEL "BIRTE" CALL SIGN DELTA ALPHA MIKE KILO
☞ POSITION SIX TWO DEGREES ONE ONE DECIMAL EIGHT MINUTES NORTH ZERO ZERO SEVEN DEGREES FOUR FOUR MINUTES EAST
☞ I HAVE PROBLEMS WITH ENGINES
☞ I REQUIRE TUG ASSISTANCE OUT

-본선은 폭발 후, 화재 발생함
-본선은 소화작업 원조를 요청함
-연기는 유독성이 아님. 오우버

2 표준 긴급 통신문

.1 구성

DSC 긴급 호출 신호를 발신한 다음, 발신기를 VHF 채널 16이나 주파수 2182kHz(자동으로 제어되지 않으면)로 전환하고 다음과 같이 긴급 통화를 개시한다:

팡 팡(3회 반복)
전 무선국(3회 반복)
여기는

- 9개의 숫자로 된 선박의 해상 이동통신업무 식별번호에 추가하여 선명/호출부호 또는 기타 식별부호
- 선박의 위치
- 긴급 통신문의 본문

.2 보기

팡-팡 팡-팡 팡-팡
전 무선국 전 무선국 전 무선국

- 여기는 2-1-1-2-3-9-6-8-0 기선 "BIRTE", 호출부호 Delta - Alpha - Mike - Kilo
- 위치는 북위 62도 11.8분 동경 007도 44분
- 본선은 주기관이 고장임
- 본선은 예인선의 원조를 요청함 아웃

3 Standard safety message

.1 Structure

After the transmission of a DSC Safety Call switch the transmitter to VHF Channel 16 or frequency 2182 kHz (if not automatically controlled) and transmit the safety message as follows:

SECURITE (repeated three times)
ALL STATIONS (or all ships in a specific geographical area, or to a specific station)(repeated three times)
THIS IS

- ☞ the 9-digit MMSI of the vessel plus name/call sign or other identification
- ☞ the text of the safety message.

.2 Example

SECURITE SECURITE SECURITE
ALL SHIPS ALL SHIPS ALL SHIPS IN AREA PETER REEF

- ☞ THIS IS
  TWO-ONE-ONE-TWO-THREE-NINE-SIX-EIGHT-ZERO
  MOTOR VESSEL "BIRTE" CALL SIGN DELTA ALPHA MIKE KILO
- ☞ DANGEROUS WRECK LOCATED IN POSITION TWO NAUTICAL MILES SOUTH OF PETER REEF OVER

3. 표준 안전 통신문

.1 구성

DSC 안전 호출신호를 발신한 다음, 발신기를 VHF채널 16이나 주파수의 2182kHz(자동으로 제어되지 않으면)로 전환하고 다음과 같이 안전 통신문을 발신한다:

씨큐리티(3회 반복)
전 무선국(또는 특정 지역에 있는 모든 선박, 또는 어떤 특정한 무선국)(3회 반복)
여기는

- 9개의 숫자로 된 해상 이동통신업무 식별번호에 추가하여 선명/ 호출부호 또는 기타 식별부호
- 안전 통신문의 본문

.2 보기

씨큐리티 씨큐리티 씨큐리티
피터 리프 해역에 있는 모든 선박 모든 선박 모든 선박

- 여기는 2-1-1-2-3-9-6-8-0 기선 "BIRTE"
  호출부호 Delta - Alpha - Mike - Kilo
- 피터 리프 남쪽 2 해리의 위치에 위험한 난파선이 확인됨 오버.

AII ON-BOARD COMMUNICATION PHRASES

AII/1 Standard wheel orders

All wheel orders given should be repeated by the helmsman and the officer of the watch should ensure that they are carried out correctly and immediately. All wheel orders should be held until countermanded. The helmsman should report immediately if the vessel does not answer the wheel. When there is concern that the helmsman is inattentive s/he should be questioned:

"What is your heading?" And s/he should respond:
"My heading is … degrees."

Order	Meaning
1. Midships	Rudder to be held in the fore and aft position.
2. Port/starboard five	5° of port/starboard rudder to be held.
3. Port/starboard ten	10°of port/starboard rudder to be held.
4. Port/starboard fifteen	15°of port/starboard rudder to be held.
5. Port/starboard twenty	20° of port/starboard rudder to be held.
6. Port/starboard twenty-five	25°of port/starboard rudder to be held.
7. Hard -a-port/starboard	Rudder to be held fully over to port/starboard.
8. Nothing to port/starboard	Avoid allowing the vessel's head to go to port/starboard
9. Meet her	Check the swing of the vessel's head in a turn.
10. Steady	Reduce swing as rapidly as possible.
11. Ease to five/ten/fifteen/ twenty	Reduce amount of rudder to 5°/10°/15°/20° and hold.
12. Steady as she goes	Steer a steady course on the compass heading indicated at the time of

AII 선내 통신 용어

AII/1 표준 조타 명령

조타수는 주어진 모든 조타 명령을 복창해야 하며 당직 사관은 조타 명령이 즉시 올바르게 수행되는지의 여부를 확인해야 한다. 모든 조타 명령은 철회될 때까지 그대로 유지되어야 한다. 타효가 없는 경우 조타수는 즉시 보고하여야 한다.

조타수가 태만할 염려가 있을 때에는 조타수에게

"선수방위는 몇 도인가?" 라고 묻고 조타수는:

"선수방위는 …도임"이라고 대답해야 한다.

명령	의미
1. Midships	키를 선수미선상에 유지하시오.
2. Port/starboard five	키를 좌/우현 5도로 유지하시오.
3. Port/starboard ten	키를 좌/우현 10도로 유지하시오.
4. Port/starboard fifteen	키를 좌/우현 15도로 유지하시오.
5. Port/starboard twenty	키를 좌/우현 20도로 유지하시오.
6. Port/starboard twenty-five	키를 좌/우현 25도로 유지하시오.
7. Hard -a-port/starboard	키를 좌/우현으로 완전히 돌리시오.
8. Nothing to port/starboard	선수를 좌/우현 쪽으로 가지 않도록 하시오.
9. Meet her	회두중, 선수의 선회를 억제하시오.
10. Steady	가능한 한 빨리 선회를 줄이시오.
11. Ease to five/ten/fifteen/twenty	타각을 5/10/15/20 도로 줄여 잡으시오.
12. Steady as she goes	명령이 내려진 순간에 컴퍼스가 지시한 안정된 선수 방위로 조타하시오.

the order. The helmsman is to repeat the order and call out the compass heading on receiving the order. When the vessel is steady on that heading, the helmsman is to call out: "Steady on …"

13. Keep the buoy/mark/beacon/…on port side/starboard side.
14. Report if she does not answer the wheel.
15. Finished with wheel, no more steering.

When the officer of the watch requires a course to be steered by compass, the direction in which s/he wants the wheel turned should be stated followed by each numeral being said separately, including zero, for example:

Order	**Course to be steered**
Port, steer one eight two	182°
Starboard, steer zero eight two	082°
Port, steer three zero five	305°

On receipt of an order to steer, for example, 182, the helmsman should repeat it and bring the vessel round steadily to the course ordered. When the vessel is steady on the course ordered, the helmsman is to call out:

"Steady on one eight two".

The person giving the order should acknowledge the helmsman's reply.

If it is desired to steer on a selected mark the helmsman should be ordered to:

"Steer on … buoy/… mark/… beacon".

The person giving the order should acknowledge the helmsman's reply.

조타수는 그 명령을 복창하고 명령을 받은 순간의 선수방위를 큰 소리로 보고한다. 선박이 그 선수 방위로 정침되면 조타수는 "…도에 정침하였음" 이라고 큰소리로 보고한다.

13. 부표/물표/비이컨/…을 좌현/우현 측에 유지하시오.
14. 타효가 없으면 보고하시오.
15. 조타 종료, 더 이상의 조타는 없음.

당직 항해사가 컴퍼스에 의하여 침로를 조타(변침)하고자 할 때에는, 타륜을 회전시킬 방향을 먼저 지시한 다음, 0을 포함한 각 숫자를 하나씩 구분하여 말해야함, 예를 들면:

명 령	조타(변침)할 침로
"Port, steer one eight two"	182°
"Starboard, steer zero eight two"	082°
"Port ,steer three zero five"	305°

조타 명령 (예를 들면 182도,)을 수령하면, 조타수는 그 명령을 복창하고 선박을 안정되게 선회시켜 명령된 침로에 정침시켜야 한다.
선박이 명령된 침로에 정침되면 조타수는:

"182도에 정침되었음"이라고 보고해야 한다.

명령을 내린 사람은 조타수의 보고를 확인해 주어야 한다.

선정된 물표를 향하여 조타하고자 할 경우에는:

"…부표/…물표/…비이콘을 향해서 조타하시오."라고 조타수에게 지시 한다.

명령을 내린 사람은 조타수의 보고를 확인해 주어야 한다.

AII/2 Standard engine orders

Any engine order given should be repeated by the person operating the bridge telegraph(s) and the officer of the watch should ensure the order is carried out correctly and immediately.[82)]

Order

1. (Port/starboard engines) Full ahead/astern
2. (Port/starboard engines) Half ahead/astern
3. (Port/starboard engines) Slow ahead/astern
4. (Port/starboard engines) Dead slow ahead/astern
5. Stop (port/starboard) engines
6. Emergency full ahead/astern
7. Stand by engine
   (Engine-room personnel fully ready to manoeuvre and bridge manned to relay engine orders.)
8. Finished with engines - no more manoeuvring.
   (Operation of engines no longer required.)

In vessels fitted with twin propellers, the word "both" should be added to all orders affecting both shafts, e.g. "Full ahead both", and "Slow astern both", except that the words "Stop all engines" should be used, when appropriate. When required to manoeuvre twin propellers indepen- dently, this should be indicated, i.e. "Full ahead starboard", "Half astern port", etc.

Where bow thrusters are used, the following orders are used:

9. Bow thruster full/half to port/starboard.
10. Stern thruster full/half to port/starboard.
11. Bow/stern thruster stop

AII/2 표준 기관 명령

선교의 텔레그래프를 작동하는 사람은 지시된 모든 기관 명령을 복창해야 하며, 당직사관은 그 명령이 올바르게 즉시 이행되는지를 확인해야 한다.

명 령

1. (좌측/우측 기관) 전/후진 전속
2. (좌측/우측 기관) 전/후진 반속
3. (좌측/우측 기관) 전/후진 미속
4. (좌측/우측 기관) 전/후진 극미속
5. (좌측/우측) 기관 정지
6. 긴급 전/후진 전속
7. 기관 사용 준비
   (기관실 요원은 기관의 사용을 위한 전원이 준비되고, 선교는 기관 명령을 전달할 인원이 배치될 것)
8. 기관 사용 종료 - 더 이상의 조종은 없음.
   (기관의 작동은 더 이상 필요 없음)

쌍축선에 있어서 두 개의 축을 모두 사용할 때는 "both(양쪽)"라는 단어를 모든 기관 명령에 덧붙여야 한다. 즉 "Full ahead both", "Slow astern both" 등. 그러나 적절한 경우 "Stop all engines"를 사용하는 것은 예외로 한다. 두 개의 프로펠러를 각각 독립적으로 사용할 필요가 있을 때에는 이 사항을 표시하여야 한다. 즉, "Full ahead starboard", "Half astern port" 등

선수 쓰러스터가 사용될 경우에는 다음과 같은 명령을 사용한다:

9. 선수 쓰러스터 좌현/우현으로 전속/반속
10. 선미 쓰러스터 좌현/우현으로 전속/반속
11. 선수/선미 쓰러스터 정지

AII/3 Pilot on the bridge

AII/3.1 Propulsion system

.1 Is the engine a diesel or a turbine?
.1.1 The engine is a diesel/turbine.
.2 Is the engine-room manned or is the engine on bridge control?
.2.1 The engine-room is manned.
.2.2 The engine is on bridge control.
.3 How long does it take to change the engines from ahead to astern?
.3.1 It takes … seconds to change the engines (from ahead to astern).
.4 How long does it take to start the engines from stopped?
.4.1 It takes … seconds to start the engines (from stopped).
.5 Is extra power available in an emergency?
.5.1 Yes, extra power is available.
.5.2 No, extra power is not available.
.6 Do you have a controllable or fixed pitch propeller?
.6.1 We have a controllable pitch propeller.
.6.2 We have a fixed pitch propeller.
.7 Do you have a right-hand or left-hand propeller?[83)]
.7.1 We have a right-hand/left-hand propeller.
.8 Do you have a single propeller or twin propellers?
.8.1 We have a single propeller/twin propellers.
.9 Do you have a bow thruster or stern thruster?
.9.1 We have one/two/… bow thruster(s)/stern thruster(s).
.10 What is the maximum manoeuvring power ahead/astern?
.10.1 The maximum manoeuvring power ahead/astern is … kilo Watts.[84)]
.11 What are the maximum revolutions ahead/astern?
.11.1 The maximum revolutions ahead/astern are ….

AII/3 도선사의 선교 승무

AII/3.1 추진기 계통

.1 기관은 디젤인가, 터빈인가?

.1.1 기관은 디젤/터빈임.

.2 기관실에는 인원이 배치되어 있는가 또는 기관은 선교 제어형인가?

.2.1 기관실에는 인원이 배치되어있음.

.2.2 기관은 선교 제어형임.

.3 기관을 전진에서 후진으로 전환하는데 시간은 얼마나 걸리는가?

.3.1 (전진에서 후진으로) 기관을 전환하는데 …초 걸림.

.4 기관을 정지 상태에서 시동하는데 시간은 얼마나 걸리는가?

.4.1 (정지 상태에서) 기관을 시동하는데 …초 걸림.

.5 비상시에 여분의 출력이 사용 가능한가?

.5.1 예, 여분의 출력이 사용 가능함.

.5.2 아니오, 여분의 출력이 사용 가능하지 않음.

.6 귀선은 가변 피치 프로펠러 또는 고정 피치 프로펠러를 장착하고 있는가?

.6.1 본선은 가변 피치 프로펠러를 장착하고 있음.

.6.2 본선은 고정 피치 프로펠러를 장착하고 있음.

.7 귀선은 우선 프로펠러 / 좌선 프로펠러를 장착하고 있는가?

.7.1 본선은 우선/좌선 프로펠러를 장착하고 있음.

.8 귀선은 단축선인가 또는 쌍축선인가?

.8.1 본선은 단축/쌍축선임.

.9 귀선은 선수 쓰러스터 / 선미 쓰러스터를 장착하고 있는가?

.9.1 본선은 한개/두개/.개의 선수/선미 쓰러스터를 장착하고있음.

.10 최대 전진/후진 조종 출력은 얼마인가?

.10.1 최대 전진/후진 조종 출력은 …킬로와트임.

.11 전진/후진 최대 회전수는 얼마인가?

.11.1 전진/후진 최대 회전수는 …임.

.12 Do the twin propellers turn inward or outward when going ahead ?[85)]

.12.1 The twin propellers turn inward/outward (when going ahead).

AII/3.2 Manoeuvring

.1 I require the pilot card/manoeuvring data.[86)]

.2 What is the diameter of the turning circle?

.2.1 The diameter of the turning circle is … metres.

.3 What is the advance and transfer distance in a crash-stop?

.3.1 The advance distance is … kilometres/nautical miles, the transfer distance is … kilometres/nautical miles,(in a crash stop).

.4 How long does it take from hard-a-port to hard-a-starboard?

.4.1 It takes … seconds (from hard-a-port to hard-a-starboard).

.5 Is the turning effect of the propeller very strong?

.5.1 Yes, the turning effect (of the propeller) is very strong.

.5.2 No, the turning effect (of the propeller) is not very strong.

.6 Where is the whistle control?

.6.1 The whistle control is on the console/on … .

.7 What notice is required to reduce from full sea speed to manoeuvring speed?[87)]

.7.1 … minutes notice is required (to reduce from full sea speed to manoeuvring speed).

.8 Do you have an automatic pilot?

.8.1 Yes, we have an automatic pilot.

.8.2 No, we do not have an automatic pilot.

.9 Give … short/prolonged blast(s) (on the whistle).

.10 Stand by lookout.

.11 Maintain a speed of … knots.

.12 What is the (manoeuvring) speed at full/half/slow/dead slow ahead?

.12 전진시 양 프로펠러는 내 선회인가 외 선회인가?
.12.1 양 프로펠러는 내/외 선회임(전진시).

AII/3.2 조종

.1 나는 파일럿 카드/조종 자료를 요구함.
.2 선회권의 직경은 얼마인가?
.2.1 선회권의 직경은 …미터임.
.3 긴급 정지시 종방향 이동 거리/횡방향 이동 거리는 얼마인가?
.3.1 (긴급 정지시) 종방향 이동 거리는 …킬로미터/해리, 횡방향 이동 거리는 …킬로미터/해리임.
.4 좌현 최대 전타에서 우현 최대 전타로 바꾸는데 얼마나 걸리는가?
.4.1 (좌현 최대전타에서 우현 최대전타로 바꾸는데) …초 걸림.
.5 프로펠러의 선회효과는 아주 강한가?
.5.1 예, (프로펠러의) 선회효과는 아주 강함.
.5.2 아니오, (프로펠러의) 선회효과는 아주 강하지 않음.
.6 기적 제어기는 어디에 있는가?
.6.1 기적 제어기는 제어탁자(콘솔)/…에 있음.
.7 항해 전속력에서 조종 속력으로 감속시키기 위해서는 어떤 통보가 필요한가?
.7.1 (항해 전속력에서 조종 속력으로 감속시키기 위해서는) … 분 통보가 필요함.
.8 귀선은 자동 조타 장치가 있는가?
.8.1 예, 본선은 자동 조타 장치가 있음.
.8.2 아니오, 본선은 자동 조타 장치가 없음.
.9 (기적) 단성/장성 …회를 울리시오.
.10 견시원을 대기시키시오.
.11 …노트의 속력을 유지하시오.
.12 전진시 (조종)전속/반속/미속/극미속의 속력은 얼마인가?

.12.1 The manoeuvring speed at full/half/slow/dead slow ahead is … knots.

.13 What is the full sea speed/fairway speed?

.13.1 The full sea speed/fairway speed is … knots.

AII/3.3 Radar

.1 Is the radar operational?

.1.1 Yes, the radar is operational.

.1.2 No, the radar is not operational.

.2 Where is the radar antenna?

.2.1 The radar antenna is on … .

.3 Does the radar have any blind sectors?

.3.1 Yes, the radar has blind sectors from … to … degrees and from … to … degrees.

.3.2 No, the radar does not have any blind sectors.

.4 Change the radar to

~ … miles range scale.

~ relative head-up/north-up/course-up.[88)]

~ true-motion north-up/course-up.

AII/ 3.4 Draft and air draft

.1 What is your present maximum draft?

.1.1 My present maximum draft is … metres.

.1.2 My draft forward/aft is … metres.

.2 What is your air draft?

.2.1 My air draft is … metres.

AII/3.5 Anchoring

.1 **Going to anchor**

.12.1 전진시 조종 전속/반속/미속/극미속의 속력은 …노트임.
.13 항해 전속력/항로 내의 제한속력은 얼마인가?
.13.1 항해 전속력/항로 내의 제한속력은 …노트임.

AII/3.3 레이더

.1 레이더는 작동 가능한가?
.1.1 예, 레이더는 작동 가능함.
.1.2 아니오, 레이더는 작동 불능임.
.2 레이더 안테나는 어디에 있는가?
.2.1 레이더 안테나는…에 있음.
.3 레이더에는 맹목구간이 있는가?
.3.1 예, 레이더에는 …도에서 …도까지 및 …도에서 …도까지 맹목구간이 있음.
.3.2 아니오, 레이더에는 맹목구간이 없음.
.4 레이더를 …마일 레인지 스케일로 바꾸시오.
레이더를 상대 운동 헤드업/노쓰업/코쓰업 지시 방식으로 바꾸시오.
레이더를 진 운동 노쓰업/코쓰업 지시 방식으로 바꾸시오.

AII/3.4 흘수 및 높이

.1 귀선의 현재의 최대 흘수는 얼마인가?
.1.1 본선의 현재의 최대 흘수는 … 미터임.
.1.2 선수/선미 흘수는 …미터임.
.2 귀선의 높이는 얼마인가?
.2.1 본선의 높이는 …미터임.

AII/3.5 묘박 (닻 정박)

.1 투묘 작업

.1 Stand by port/starboard/both anchor(s) for letting go.
.2 Walk out the anchor(s).
.3 We are going to anchorage.
.4 We will let go port/starboard/both anchor(s).
.5 Put … shackles in the water/in the pipe/on deck.
.6 Walk back port/starboard/both anchor(s) one/one and
a half shackle(s).
.7 We will let go port/starboard/both anchor(s) … shackle(s) and dredge it/them.
.8 Let go port/starboard/both anchor(s).
.9 Slack out the cable(s).
.9.1 Check the cable(s).
.9.2 Hold on the port/the starboard/both cable(s).
.10 How is the cable leading?
.10.1 The cable is leading
~ ahead/astern.
~ to port/to starboard.
~ round the bow.
~ up and down.[89)]
.11 How is the cable growing?
.11.1 The cable is slack/tight/coming tight.
.12 Is/are the anchor(s) holding?
.12.1 Yes, the anchor(s) is/are holding.
.12.2 No, the anchor(s) is/are not holding.
.13 Is she brought up?
.13.1 Yes, she is brought up in position … .[90)]. **(p.351 참조)**
.13.2 No, she is not brought up (yet).
.14 Switch on the anchor light(s).
.15 Hoist the anchor ball.[91)]
.16 Check the anchor position by bearings/by … .
.16.1 The anchor position is bearing … degrees, distance …. kilometres/nautical miles to ….

.1 좌현/우현/양현 묘를 투묘할 준비를 하시오.
.2 닻을 풀어서 투묘 준비 상태로 하시오.
.3 본선은 투묘지로 가고 있음.
.4 본선은 좌현/우현/양현 묘를 투묘할 것임.
.5 닻줄 …절(샤클)을 수면/묘쇄공/갑판상까지 풀어 주시오.
.6 좌현/우현/양현 닻줄을 1/1.5 절(샤클)까지 (양묘기를 역전 시켜) 풀어 주시오.
.7 본선은 좌현/우현/양현묘를 …절로 투묘하고 그것/그것들을 끌겠음.
.8 좌현/우현/양현 묘를 투묘하시오.
.9 닻줄을 늦추어 주시오.
.9.1 닻줄을 억제하시오.
.9.2 좌현/우현/양현 닻줄을 멈춘 채로 있으시오.
.10 닻줄 방향은 어떤가?
.10.1 닻줄 방향은 정선수/정선미임
닻줄 방향은 좌현/우현 측임
닻줄 방향은 선수를 감아 돌았음.
닻줄 방향은 수직임.

.11 닻줄의 장력은 어떻게 돼가고 있는가?
.11.1 닻줄은 느슨함/팽팽함/팽팽해지고 있음
.12 묘/양현 묘는 파주력이 있는가?
.12.1 예, 닻(들)은 파주력이 있음.
.12.2 아니오, 닻(들)은 파주력이 없음.
.13 본선의 닻은 해저에 완전히 박혔는가?
.13.1 예, 닻은 위치 …에서 해저에 완전히 박혔음.
.13.2 아니오, 닻은 해저에 (아직)완전히 박히지 않았음.
.14 정박등을 켜시오.
.15 흑구를 게양하시오.
.16 방위를 측정하여/…하여 투묘 위치를 점검하시오.
.16.1 투묘 위치는 …를 향하여 방위…도, 거리…킬로미터/해리임.

.16.2 Check the anchor position every … minutes.

.2 Leaving the anchorage

.1 How much cable is out?

.1.1 … shackle(s) is/are out.

.2 Stand by for heaving up.

.3 Put the windlass in gear.

.3.1 The windlass is in gear.

.4 How is the cable leading?

.4.1 The cable is leading

~ ahead/astern.

~ to port/to starboard.

~ round the bow.

~ up and down.

.5 Heave up port/starboard/both cable(s).

.6 How much weight is on the cable?

.6.1 Much/too much weight is on the cable.

.6.2 No weight is on the cable.

.7 Stop heaving.

.8 How many shackles are left (to come in)?

.8.1 … shackles are left (to come in).

.9 Attention! Turn in cable(s).

.10 The anchor(s) is/are aweigh.

.10.1 The cables are clear.

.11 The anchor(s) is/are clear of the water/home/foul/ secured.

AII/3.6 Tug assistance

.1 We will take … tug(s).

.2 The tug(s) will pull/push.

.3 We use the towing line(s) of your vessel.

.16.2　매 …분마다 투묘 위치를 점검하시오.

.2　투묘지 출발

.1　닻줄은 몇 절(샤클) 풀려 나가 있는가?
.1.1　…절(샤클) 풀려 나가 있음.
.2　감아 올릴 준비를 하시오.
.3　양묘기의 기어를 연결하시오.
.3.1　양묘기의 기어는 연결되었음.
.4　닻줄 방향은 어떠한가?
.4.1　닻줄 방향은 정선수/정선미임
닻줄 방향은 좌현/우현 측임
닻줄 방향은 선수를 감아 돌았음.
닻줄 방향은 수직임.
.5　좌현/우현/양현 닻줄을 감아들이시오.
.6　닻줄에는 중량이 얼마나 걸려있는가?
.6.1　닻줄에는 중량이 많이/너무 많이 걸려있음.
.6.2　닻줄에는 중량이 걸려 있지 않음.
.7　감아 올리기를 멈추시오.
.8　(더 감아들여야 할)닻줄은 몇 절(샤클)이나 남아 있는가?
.8.1　(더 감아들여야 할) …절(샤클)이 남아 있음.
.9　주의하시오! 닻줄이 꼬였음.
.10　닻(들)은 발묘되었음.
.10.1　닻줄은 엉켜있지 않음.
.11　닻(들)은 수면위로 떠올랐음/묘쇄 공으로 들어왔음/엉켜 있음/완전히 격납되었음.

AII/3.6 예인선의 원조

.1　본선은 …척의 예인선을 사용하겠음.
.2　예인선은 당기겠음/밀겠음.
.3　우리는 귀선의 예인줄을 사용하겠음.

.3.1 We use the towing line(s) of the tug(s).
.4 Stand by for making fast the tug(s).
.5 Use the centre lead/panama lead.[92]
.5.1 Use the fairlead
~ on port side/starboard side.
~ amidships.
~ on port bow/starboard bow.
~ on port/starboard quarter.
.6 Send heaving line(s) to the tug(s).
.7 Send two towing line(s) to the tug(s).
.8 Lower towing line(s)
~ to the tug(s).
~ … metre(s) from the water.
.9 Slack away towing line(s).
.10 Make fast the tug(s).
.10.1 Make fast the tug(s)
~ forward/aft.
~ on port bow/starboard bow.
~ on port quarter/starboard quarter.
.11 Make fast the forward/aft tug(s) alongside on port side/ starboard side.
.12 Make fast … tug(s) on each bow/quarter.
.13 Put the eyes of the towing line(s) on bitts.[93]
.14 The tug(s) is/are fast (on …).
.15 Keep clear of towing line(s).
.16 Stand by for letting go the tug(s).
.17 Let go the tug(s).
.18 Towing line(s) is/are broken.

AII/3.7 Berthing and unberthing

.1 General

.3.1 우리는 예인선의 예인줄을 사용하겠음.
.4 예인선을 잡아 맬 준비를 하시오.
.5 중앙/파나마 도삭기를 사용하시오.
.5.1 좌현/우현 측 도삭기를 사용하시오.
선체 중앙부의 도삭기를 사용하시오.
좌현 선수/우현 선수의 도삭기를 사용하시오.
좌현/우현 선미의 도삭기를 사용하시오.
.6 예인선에 히빙 라인을 보내시오.
.7 예인선에 두 개의 예인줄을 보내시오.
.8 예인선에 예인줄을 내려보내시오.
수면상 …미터까지 예인줄을 내려보내시오.
.9 예인줄을 늦추어 주시오.
.10 예인선을 잡아매시오.
.10.1 선수/선미에 예인선을 잡아매시오.
좌현 선수/우현 선수에 예인선을 잡아매시오.
좌현 선미/우현 선미에 예인선을 잡아매시오.
.11 선수/선미쪽의 예인선은 좌현/우현 측에 나란히 잡아매시오.
.12 양쪽 선수/선미에 …척 씩의 예인선을 잡아매시오.
.13 예인줄의 고리를 계주에 거시오.
.14 예인선(들)은 (…에) 잡아매었음.
.15 예인줄으로부터 떨어지시오.
.16 예인선을 풀어줄 준비를 하시오.
.17 예인선을 풀어주시오.
.18 예인줄(들)이 절단되었음.

AII/3.7 접안 및 이안

.1 일반 사항

.1 Is/are the propeller(s) clear?
.1.1 Yes, the propeller(s) is clear.
.1.2 No, the propeller(s) is not clear.
.1.3 Keep the propeller(s) clear.
.2 Are fenders on the berth?
.2.1 Yes, fenders are on the berth.
.2.2 No, fenders are not on the berth.
.3 Have fenders ready fore and aft.

.2 Berthing[94)]

.1 We will berth port side/starboard side alongside.
.2 We will moor
~ to buoy(s) (ahead and astern).
~ alongside.
~ to dolphins.
.3 Send out
~ the head/stern/breast lines.
~ the … spring(s) forward/aft.
.4 Do you have tension winches?
.4.1 Yes, we have tension winches (forward and aft).
.4.2 No, we do not have tension winches.
.5 Have the heaving lines ready forward and aft.
.6 Send the heaving/head/stern/breast line(s) ashore.
.7 The linesmen will use shackles/lashings for securing the mooring.
.8 Use
~ the centre lead/panama lead.
~ the bow lead.
~ the port quarter/starboard quarter lead.
.9 Heave on the … line(s)/… spring(s).
.10 Pick up the slack on the … line(s)/… spring(s).
.11 Heave away.
.11.1 Stop heaving.

.1 프로펠러는 자유로이 움직일 수 있는가(장애물이 없는가)?
.1.1 예, 프로펠러는 자유로이 움직일 수 있음(장애물이 없음).
.1.2 아니오, 프로펠러는 자유로이 움직일 수 없음(장애물 있음).
.1.3 프로펠러를 자유로이 움직일 수 있도록(장애물이 없도록) 하시오.
.2 방현재는 계류지에 있는가?
.2.1 예, 방현재는 계류지에 있음.
.2.2 아니오, 방현재는 계류지에 있지 않음.
.3 선수, 선미에 방현재를 준비하시오.

.2 접안

.1 본선은 좌현/우현을 안벽에 대고 접안하겠음.
.2 본선은 (선수, 선미를) 부표에 계류하겠음.
본선은 옆에 대고 계류하겠음.
본선은 돌핀에 계류하겠음.
.3 선수/선미/옆 줄(계류삭)을 내어주시오.
선수/선미 쪽의 스프링 …개를 내어주시오.
.4 귀선은 장력 윈치를 장착하고 있는가?
.4.1 예, 본선은 (선수 및 선미에) 장력 윈치를 장착하고 있음.
.4.2 아니오, 본선은 장력 윈치를 장착하고 있지 않음.
.5 선수, 선미에 히빙 라인을 준비하시오.
.6 히빙라인/선수줄/선미줄/옆줄을 육상으로 보내시오.
.7 줄잡이는 계류삭을 고정시키기 위하여 연결용 고리/결박용 밧줄을 사용할 것임.
.8 중앙부/파나마 도삭기를 사용하시오.
선수 도삭기를 사용하시오.
좌현 선미/우현 선미의 도삭기를 사용하시오.
.9 …줄/…스프링을 계속 감으시오.
.10 …줄/…스프링의 늘어짐을 팽팽하게 하시오.
.11 계속 감아들이시오.
.11.1 감아들이기를 멈추시오.

.12 Slack away/check the … line(s)/… spring(s).
.13 Hold on the … line(s)/… spring(s).
.14 Heave in easy.
.14.1 Heave alongside.
.15 Keep the … line(s)/… spring(s) tight.
.16 Report the forward/aft distance to … .
.16.1 The forward/aft distance to … is … metres.
.17 We have to move … metres ahead/astern.
.18 We are in position.
.19 Make fast fore and aft.
.20 Finished with manoeuvring stations.

.3 Unberthing

.1 Stand by engine(s).
.2 Are you ready to get underway?
.2.1 Yes, we are ready (to get underway).
.2.2 No, we are not ready (yet) (to get underway).
.2.3 We will be ready to get underway in … minutes.
.3 Stand by for letting go.
.4 Single up the … lines and … springs fore and aft.
.5 Slack away/hold on/heave on the
~ head/stern line.
~ breast line.
~ fore/aft spring.
.6 Let go
~ the head/stern line.
~ the breast line.
~ the fore/aft spring.
~ all (forward/aft).
.7 Let go the towing line(s).
.8 Stand by bow anchor(s).
.9 Finished with manoeuvring stations.

.12 …줄/…스프링을 계속 늦춰 주시오/억제하시오.
.13 …줄/…스프링을 그 자리에 붙잡으시오.
.14 천천히 감아들이시오.
.14.1 안벽 쪽으로 끌어당기시오.
.15 …줄/…스프링을 팽팽하게 유지하시오.
.16 선수/선미와 …사이의 거리를 보고하시오.
.16.1 선수/선미와 …사이의 거리는 …미터임.
.17 본선은 전방/후방으로 …미터 이동하여야 함.
.18 본선은 정위치임.
.19 선수, 선미를 잡아매시오.
.20 조종부서를 종료함.

.3 이안

.1 기관 사용 준비를 하시오.
.2 귀선은 항해를 개시할 준비가 되었는가?
.2.1 예, 본선은 (항해를 개시할)준비가 되었음.
.2.2 아니오, 본선은 (항해를 개시할)준비가 (아직) 안되었음.
.2.3 본선은 …분 후 항해를 개시할 준비가 됨.
.3 (계류삭을) 풀어줄 준비를 하시오.
.4 선수, 선미에 …줄과 …스프링을 싱글 업하시오.
.5 선수/선미 줄을 계속 늦춰주시오/그 자리에 붙잡으시오/계속 감으시오.
옆줄을 계속 늦춰주시오/그자리에 붙잡으시오./계속 감으시오.
선수/선미 스프링을 계속 늦춰주시오/그 자리에 붙잡으시오./계속 감으시오.
.6 선수/선미 줄을 풀어주시오.
옆줄을 풀어주시오.
선수/선미 스프링을 풀어주시오.
(선수/선미의) 모든 계류삭을 풀어주시오.
.7 예인줄을 풀어주시오.
.8 선수묘를 준비하시오.
.9 조종부서를 종료하였음.

STANDARD MARINE COMMUNICATION PHRASES: PART B

PART B covers further on-board standard safety-related phrases which, supplementary to PART A, may assist mariners in meeting other basic on-board communication requirements and may be regarded useful for Maritime English instruction.

B ON-BOARD COMMUNICATION PHRASES

B1 Operative shiphandling

B1/1 Handing over the watch

B1/1.1 Briefing on position, movement and draft

The officer of the watch should brief the relieving officer on the following:

.1 Position

.1 The present position is
~ latitude … , longitude … .
~ bearing … degrees, distance … cables/nautical miles.
~ buoy …(*charted name*).
~ between … and… .
~ way point/reporting point … .
~ … .

.2 The next way point/reporting point is … .

.3 ETA at … is … UTC.

.4 We are passing/we passed buoy …(*charted name*) on port side/starboard side.

표준 해사 통신 영어: B편

B편은, A편에 부가하여, 추가의 선내 안전 관련 표준 용어들을 포함하며, 항해자들로 하여금 다른 기초적인 선내 통신 요구 사항을 충족하도록 도울 것이며, 해사 영어교육에 유용하게 여겨질 것이다.

B 선내 통신 용어

B1 선박 운항

B1/1 당직의 인계

B1/1.1 선위, 이동, 흘수에 대한 상황 설명
당직 사관은 다음 사항에 대하여 인계 받는 사관에게 설명해야한다:

.1 **선위**

.1 현재 위치는 위도… , 경도…임.
현재 위치는 방위…도, 거리…케이블/해리임.
현재 위치는 …(해도에 기재된 명칭) 부표임.
현재 위치는 …와 …사이임.
현재 위치는 통과 지점/보고 지점…임.
현재 위치는 …임.

.2 다음의 통과 지점/보고 지점은…임.

.3 …에의 도착 예정 시각은 세계시로…시임.

.4 본선은 …(해도에 기재된 명칭) 부표를 좌현/우현 측으로 통과하고 있음/통과하였음.

.5 We are approaching buoy …(*charted name*) on port side/ starboard side.

.6 Buoy …(*charted name*) is … cables/nautical miles ahead.

.7 We are entering/we entered area … .

.8 We are leaving/we left area … .

.2 Movements

.1 True course/gyro compass course/magnetic compass course is … degrees.

.2 Gyro compass error is … degrees plus/minus.

.2.1 Magnetic compass error is … degrees east/west.

.3 Speed over ground/through water is … knots.

.4 Set and drift is … degrees, … knots.

.5 We are making … degrees leeway.

.6 The course board is written up.

.7 The next chart is within …. hours.

.3 Draft

.1 Draft forward/aft is … metres.

.2 Present maximum draft is … metres.

.3 Underkeel clearance is … metres.

B1/1.2 Briefing on traffic situation in the area

.1 A vessel is

~ overtaking …*(cardinal points)* of us.

~ on opposite course.

~ passing on port side/starboard side.

.2 A vessel is crossing from port side.

.5 본선은…(해도에 기재된 명칭) 부표를 좌현/우현 측으로 접근하고 있음.

.6 …(해도에 기재된 명칭) 부표는 …케이블/해리 전방에 있음.

.7 본선은…해역에 진입하고 있음/진입하였음.

.8 본선은…해역을 떠나고 있음/ 떠났음.

.2 이동

.1 진/자이로컴파스/자기컴파스 침로는 …도임.

.2 자이로 컴파스의 오차는 …도 플러스/마이너스임.

.2.1 자기 컴파스의 오차는 …도 편동/편서임.

.3 대지/대수 속력은 …노트임.

.4 유향과 유정은 …도 방향, …노트임.

.5 본선은…도의 풍압 편위가 생기고 있음.

.6 침로 기재판이 기재되었음.

.7 …시간 이내에 다음 해도를 사용함.

.3 흘수

.1 선수/선미 흘수는 …미터임.

.2 현재 최대 흘수는…미터임.

.3 선저 여유 수심은 …미터임.

B1/1.2 당해 해역의 통항 상태에 대한 상황 설명

.1 선박이 본선의…(방점)측을 추월하고 있음.
반대 침로로 가고 있는 선박이 있음.
좌현/우현 측에 통과선이 있음.

.2 좌현측으로부터 횡단하는 선박이 있음.

.2.1 The vessel
~ will give way.
~ has given way.
~ has not given way yet.
~ is standing on.
~ need not give way.

.3 A vessel is crossing from starboard side.

.3.1 We
~ need not give way.
~ will stand on.
~ will alter course to give way.
~ have altered course to give way.

.3.2 The vessel will pass … kilometres/nautical miles ahead/astern.

.3.3 I will complete the manoeuvre.

.4 A vessel …*(cardinal points)* of us is on the same course.

.5 The bearing to the vessel in … degrees is constant.

.6 There is heavy traffic/… in the area.

.6.1 There are fishing boats/… in the area.

.7 There are no dangerous targets on the radar.

.7.1 Attention. There are dangerous targets on the radar.

.8 Call the Master if any vessel passes with a CPA of less than … miles.

.8.1 Call the Master if … .

B1/1.3 Briefing on navigational aids and equipment status

.1 Port side/starboard side radar is at … miles range scale.

.2 The radar is
~ relative head-up/north-up/course-up.
~ true-motion north-up/course-up.

.3 GPS/LORAN is/is not in operation.

.4 Echo sounder is at … metres range scale.

.2.1 상대선이 피항할 것임.
상대선이 피항하였음.
상대선이 아직 피항하지 않았음.
상대선은 (상태를) 유지하고 있음.
상대선은 피항할 필요가 없음.
.3 선박이 우현 측으로부터 횡단하고 있음.
.3.1 본선은 피항할 필요가 없음.
본선은 (상태를)유지할 것임.
본선은 피항하기 위하여 변침할 것임.
본선은 피항하기 위하여 변침하였음.
.3.2 상대선은 선수/선미 방향 …킬로미터/해리를 통과할 것임.
.3.3 본인이 조종을 완료할 것임.
.4 본선의 …(방점)쪽에 있는 선박은 같은 침로임.
.5 …도 방향에 있는 상대선의 방위는 변치 않고 있음.
.6 이 해역에 많은 통항선/…이 있음.
.6.1 이 해역에 어선들/…이 있음.
.7 레이더 상에 위험한 표적이 없음.
.7.1 주의. 레이더 상에 위험한 표적이 있음.
.8 어떤 선박이 CPA …마일 이내로 통과하면 선장을 호출 하시오.
.8.1 만일 …이면 선장을 호출하시오.

B1/1.3 항행 원조 장치와 장비 상태에 대한 상황 설명

.1 좌현/우현 측 레이더는 … 마일 레인지 스케일임.
.2 레이더는 상대 운동 헤드업/노쓰업/코스업 지시 방식임.
레이더는 진운동 노쓰업/코스업 지시 방식임.
.3 GPS/LORAN은 작동 중임./작동중이 아님.
.4 음향 측심의는 …미터 레인지 스케일임.

.4.1 The echo sounder recordings are unreliable.
.5 I changed to manual/automatic steering (at ··· UTC).
.6 Navigation lights are switched on/off.

B1/1.4 Briefing on radiocommunications

.1 INMARSAT ···(*type of system*) is operational/is not operational.
.2 VHF DSC Channel 70/VHF Channel ···/DSC controller is switched on.
.2.1 DSC frequency 2187.5 kHz is switched on.
.3 NAVTEX is switched on.[95)]
.4 Following was received on ··· at ··· UTC
.5 Shore based radar assistance/VTS/Pilot Station is on VHF Channel ··· .
.6 The Pilot Station/VTS station requires
~ flag state.
~ call sign/identification.
~ draft.
~ gross tonnage.
~ length overall.
~ kind of cargo.
~ ETA at ··· .
~ MAREP POSREP/··· .[96)]
~ ··· .

B1/1.5 Briefing on meteorological conditions

.1 A weak/strong (tidal) current is setting ··· degrees.
.1.1 The direction of the (tidal) current will change in ··· hours.
.2 Fog/mist/dust/rain/snow/··· is in the area.
.3 Automatic fog signal is switched on.
.4 The wind increased/decreased (within last ··· hours).

.4.1 음향 측심의의 기록은 신뢰할 수 없음.
.5 본인이(세계시…시에) 수동/자동 조타로 전환하였음.
.6 항해등은 점등/소등하였음.

B1/1.4 무선 통신에 대한 상황 설명

.1 INMARSAT …(시스템 형식) 이 작동 가능함/작동 불능임.
.2 VHF DSC 채널 70번/VHF 채널 …/DSC 제어기는 켜졌음.
.2.1 DSC 주파수 2187.5 kHz는 켜졌음.
.3 NAVTEX는 켜졌음.
.4 세계시…시에 …으로 다음 사항을 수신하였음
.5 VHF 채널 …에서 해안 레이더 원조/VTS/파일럿 스테이션과 교신 중임.
.6 파일럿 스테이션/ VTS 국이 선적국명을 묻고 있음.
파일럿 스테이션/VTS 국이 호출부호/식별부호를 묻고 있음.
파일럿 스테이션/ VTS 국이 흘수를 묻고 있음.
파일럿 스테이션/ VTS 국이 총톤수를 묻고 있음.
파일럿 스테이션/ VTS 국이 전장을 묻고 있음.
파일럿 스테이션/ VTS 국이 화물 종류를 묻고 있음.
파일럿 스테이션/ VTS 국이 …에의 ETA를 묻고 있음.
파일럿 스테이션/ VTS 국이 MAREP POSREP/…을 묻고 있음.
파일럿 스테이션/ VTS 국이 …을 묻고 있음.

B1/1.5 기상 상태에 관한 상황 설명.

.1 약한/강한 (조류)해류가 …도 방향으로 흐르고 있음.
.1.1 (조류)해류의 방향은 …시간 이후에 변할 것임.
.2 이 해역에 안개/엷은 안개/먼지/강우/강설/…이 있음.
.3 자동 무중신호기가 켜졌음.
.4 (지난 …시간 동안에) 바람이 증가하였음/감소하였음.

.4.1 The wind is …(*cardinal points*) force Beaufort … .
.4.2 The wind changed from …(*cardinal points*) to …(*cardinal points*).
.5 The sea state is expected to change (within … hours).
.6 A smooth/moderate/rough/heavy sea slight/moderate/high swell of … metres from …(*cardinal points*) is expected (within … hours).
.7 A tsunami/an abnormal wave is expected by … UTC.
.8 Visibility is … nautical miles.
.9 Visibility is reduced by fog/mist/dust/rain/snow/… .
.10 Visibility is expected
~ to decrease/increase to … nautical miles (within … hours).
~ variable between … and … nautical miles (within … hours).
.11 Next weather report is at … UTC.
.12 Atmospheric pressure is … millibars/hPascal.
.13 Barometric change is … millibars/hPascal per hour/ within the last … hours.
.13.1 Barometer is steady/dropping (rapidly)/rising (rapidly).
.14 There was a gale warning/tropical storm warning for the area … at … UTC.

B1/1.6 Briefing on standing orders and bridge organization

.1 Standing orders for the period from … to … UTC are: … .
.2 Standing orders for the area … are : … .
.3 Take notice of changes in the standing orders.
.4 Do you understand the standing orders?[97)]
.4.1 Yes, I understand the standing orders.
.4.2 No, I do not understand, please explain.
.5 Read/sign the standing orders.
.6 The latest fire patrol was at … UTC.

.4.1 풍향은 …(방점), 풍력계급은 …임.
.4.2 풍향은 …(방점)에서 …(방점)으로 변하였음.
.5 (…시간 이내에) 해면 상태는 변할 것으로 예상됨.
.6 …(방점)방향, 파고…미터의 잔잔한/보통인/거친/높은 파도/ 약간인/보통인/심한 너울이 (…시간 이내에) 예상됨.
.7 세계시 …시경까지 쓰나미/비정상적인 파도가 예상됨.
.8 시정은 …해리임.
.9 시정은 안개/엷은 안개/먼지/강우/강설/ …로 인하여 감소하였음.
.10 시정은(…시간 이내에)…해리로 감소/증가할 것으로 예상됨.
시정은 (…시간 이내에) …해리에서 …해리 사이에 가변적일 것으로 예상됨.
.11 다음 일기 예보는 세계시…시에 있음.
.12 기압은 …밀리바/헥토파스칼임.
.13 기압의 변화율은 매 시간당/지난…시간 동안…밀리바/헥토파스칼임.
.13.1 기압은 변함없음/(급격히)하강하고 있음/(급격히)상승하고 있음.
.14 세계시…시에 …해역에 강풍 경보/열대성 폭풍 경보가 발령되었음.

B1/1.6 당직 근무 수칙, 선교 조직에 대한 상황 설명

.1 세계시…시부터…시까지의 기간에 대한 당직근무수칙은:…임.
.2 …해역에 대한 당직 근무 수칙은: …임.
.3 당직 근무 수칙의 변경 사항에 주목하시오.
.4 당신은 당직 근무 수칙을 이해하는가?
.4.1 예, 나는 당직 근무 수칙을 이해함.
.4.2 아니오, 나는 이해하지 못함, 설명하여 주시오.
.5 당직 근무 수칙을 읽으시오/서명하시오.
.6 최근의 화재 순찰은 세계시…시에 행하였음.

.7 The latest security patrol was at ··· UTC.
.7.1 Everything is in order.
.7.2 The following was stated: ··· .
.7.3 The following measures were taken: ··· .
.7.4 The following requires attention: ··· .
.8 The lookout is standing by.
.9 The helmsman is standing by.
.10 Call the Master at ··· UTC/in position ··· .

B1/1.7 Briefing on special navigational events

Also see AI/3 "Safety communications".

.1 There was an engine alarm at ··· UTC due to ··· .
.2 Speed was reduced at ··· UTC due to ··· .
.3 Engine(s) was/were stopped at ··· UTC due to ··· .
.4 Course was altered at ··· UTC due to ··· .
.5 The Master/Chief Engineer was called at ··· UTC due to ··· .

B1/1.8 Briefing on temperatures, pressures and soundings

.1 The ···*(equipment)* temperature minimum/maximum is
~ ··· degrees (centigrade) /to maintain.
~ ··· above/below normal.
~ critical.
.1.1 Do not exceed a minimum/maximum temperature of ··· degrees.
.2 The ···*(equipment)* pressure minimum/maximum is
~ ··· kilopounds/bars/to maintain.
~ above/below normal.
~ critical.
.2.1 Do not exceed a pressure of ··· kilopounds/bars.
.3 Ballast/fresh water/ fuel/oil /slop sounding is ··· metres/ cubic metres.

.7 최근의 안전 순찰은 세계시…시에 행하였음.
.7.1 모든 것은 정상임.
.7.2 다음 사항이 언급되었음: … .
.7.3 다음의 조치를 취하였음: … .
.7.4 다음 사항에 주의가 요구됨: … .
.8 견시원이 대기 중임.
.9 조타수가 대기 중임.
.10 세계시 …시에/위치 …에서 선장을 호출하시오.

B1/1.7 특기 사항에 대한 상황 설명.

(AI/3의 "안전 통신"편도 참조할 것)

.1 세계시 …시에 …로 인하여 기관 경보가 울렸음.
.2 세계시 …시에 …로 인하여 감속하였음.
.3 세계시 …시에 …로 인하여 기관을 정지하였음.
.4 세계시 …시에 …로 인하여 변침하였음.
.5 세계시 …시에 …로 인하여 선장/기관장이 호출되었음.

B1/1.8 온도, 압력, 측심에 대한 상황 설명

.1 …(장비) 의 최저/최고 온도는 (섭씨)…도임/유지해야 함.
…(장비) 의 최저/최고 온도는 정상치보다 높음/ 낮음.
…(장비) 의 최저/최고 온도는 임계치임.
.1.1 최고/최저 온도 …도를 넘지 마시오.
.2 …(장비)의 최소/ 최고 압력은 …킬로 파운드임/ 바아임/로 유지되어야 함.
…(장비)의 최소/ 최고 압력은 정상치보다 높음/ 낮음.
…(장비)의 최소/ 최고 압력은 임계치임.
.2.1 …킬로 파운드/ 바아의 압력을 초과하지 마시오.
.3 밸러스트/ 청수/ 연료/ 유류/ 슬롭의 측심 결과는 …미터/입방 미터임.

.3.1 Sounding of
~ no. ··· cargo tank is ··· metres/cubic metres.
~ no. ··· cargo hold is ··· centimetres.
~ ··· is ··· .

B1/1.9 Briefing on operation of main engine and auxiliary equipment

Also see B1/1.8.

.1 (present) revolutions of the main engine(s) are ··· per minute.
.2 (present) output of the main engine(s)/auxiliary engine(s) are ··· kilowatts.
.3 (present) pitch of the propeller(s) is ··· degrees.
.4 There are no problems.
.5 There are problems
~ with the main engine(s)/auxiliary engine(s).
~ with ··· .
.6 Call the watch engineer (if the problems continue).
.6.1 Call the watch engineer ··· minutes before the arrival at ···/at ···UTC.

B1/1.10 Briefing on pumping of fuel, ballast water, etc.

.1 There is no pumping at present.
.2 We are filling/we filled(no.) ··· double bottom tank(s) /the ballast tanks/the ··· tank(s).
.2.1 Fill up ··· tonnes/sounding ···/ullage ···/level ··· to the alarm point.
.3 We are discharging/we discharged(no.) ··· double bottom tank(s)/the ballast tanks/the ··· tank(s).
.4 We are transferring/we transferred fuel/ballast/fresh water/oil from (no.) ··· tank(s) to (no.) ··· tank(s).
.5 We require a further generator to operate an additional pump.

.3.1 제…번 화물 탱크의 측심 결과는 …미터/입방미터임.
제…번 화물창의 측심 결과는 …센티미터임.
…의 측심 결과는 …임.

B1/1.9 주기관 및 보조 설비의 운전에 대한 상황 설명

(B1/1.8참조)

.1 (현재의) 주기관 회전수는 매분당 …임.
.2 (현재의) 주기관/보조 기관의 출력은 …킬로와트임.
.3 (현재의) 프로펠러의 피치는 …도임.
.4 문제거리가 없음.
.5 주기관/보조기관에 문제가 있음.
…에 문제가 있음.
.6 (문제가 계속되면) 당직 기관사에게 전화하시오.
.6.1 …에 도착하기 …분 전에/세계시 …시에 당직 기관사에게 전화하시오.

B1/1.10 연료, 밸러스트 등의 이송에 대한 상황 설명

.1 현재 펌핑은 실시하지 않음.
.2 우리는 (번호)…번 이중저 탱크/ 밸러스트 탱크/ …탱크를 채우고 있음/ 채웠음.
.2.1 경보점까지 …톤을/ 깊이…로/ 얼레지…까지/ 수위…로 채우시오.
.3 우리는 (번호)…번 이중저탱크/ 밸러스트탱크/ … 탱크를 배출하고 있음/ 배출하였음.
.4 우리는 …번 탱크에서 …번 탱크로 연료/ 밸러스트/ 청수/ 유류를 이송하고 있음/이송하였음.
.5 우리는 추가의 펌프를 작동하기 위하여 발전기를 더 필요로 함.

B1/1.11 Briefing on special machinery events

.1 There was a breakdown of the main engine(s) (at … UTC/from … to … UTC).
.1.1 There was a breakdown of … (at … UTC/from … to … UTC).
.2 There was a total blackout (at … UTC/from … to … UTC).
.2.1 There was a blackout in …(at … UTC/from … to … UTC).
.3 Main engine(s) was/were stopped (at … UTC/from … to …UTC) due to … .
.4 Speed was reduced (at … UTC/from … to … UTC) due to … .
.5 Call the Master/Chief engineer if the revolutions of the main engine(s) are below … per minute.
.5.1 Call the Master/Chief Engineer/watch engineer if … .

B1/1.12 Briefing on record keeping

.1 The log books/record books are completed and signed.
.1.1 The note book entries will be copied (into the log books/record books) after the watch.
.2 Change the paper of the data logger/echo sounder/… recorder.
.2.1 Refill the toner/ink of the data logger/echo sounder/… recorder

B1/1.13 Handing and taking over the watch

The Master/Chief Engineer or an (engineer) officer handing over the watch should say:

.1 You have the watch now.

The relieving officer should confirm and say:

.1.1 I have the watch now .

B1/1.11 기관 특기 사항에 대한 상황 설명

.1 (세계시…시에/세계시…시부터…시까지) 주기관이 고장 났음.
.1.1 (세계시…시에/세계시…시부터…시까지)…이 고장 났음.
.2 (세계시…시에/세계시…시부터…시까지) 전부 정전임.
.2.1 (세계시…시에/세계시…시부터…시까지) …에 정전임.
.3 (세계시…시에/세계시…시부터…시까지) …로 인하여 주기관이 정지되었음.
.4 (세계시…시에/세계시…시부터…시까지)…로 인하여 감속 하였음.
.5 주기관의 회전수가 매분당 …미만이 되면 선장/기관장을 호출 하시오.
.5.1 만약 …이면 선장/기관장/당직 기관사를 호출하시오.

B1/1.12 기록 유지에 대한 상황 설명

.1 항해일지/ 기록부가 완성되고 서명되었음.
.1.1 노우트의 기재사항은 당직 후(항해일지/기록부)에 옮겨 적을 것임.
.2 데이터 로거/ 음향 측심의/ …기록기의 기록지를 바꾸시오.
.2.1 데이터 로거/ 음향 측심의/ …기록기의 토너/잉크를 재충전 하시오.

B1/1.13 당직의 인계·인수

당직을 인계하는 선장/기관장이나 항해사(기관사)는 다음과 같이 말하여야 함:
.1 당신은 지금 당직을 인계 받음.

당직을 인수하는 항해사는 확인하고 다음과 같이 말해야 함:
.1.1 나는 지금 당직을 인수함.

The Master/Chief Engineer when called to the bridge /engine (control) room and formally taking over the watch, should confirm and say:

.2 I have the watch now.

The officer of the watch should confirm and say:

.2.1 You have the watch now.

B1/2 Trim, list and stability

.1 The vessel is on even keel (at present).

.1.1 The vessel is … metres down by the head/stern (at present).

.2 There is no list (at present).

.2.1 (Present) list is … degrees to port/starboard.

.3 Fuel/ballast/fresh water/oil was transferred from (no.)… tank(s) to (no.)… tank(s) to correct the list.

.3.1 We must transfer fuel/ballast/fresh water/oil from (no.)… tank(s) to (no.)… tank(s) to correct the list.

.4 Deck cargo/cargo (in (no.) … hold(s)) was restowed to correct the list.

.4.1 We must restow cargo in (no.) … hold(s) to correct the list.

.5 (Present) stability is good/poor.

.6 (no.) … double bottom tank(s) was/were filled to improve the stability.

.7 Fuel/ballast/fresh water/oil was transferred from (no.)… tank(s) to (no.) … tank(s) to improve the stability.

.7.1 We must transfer fuel/ballast/fresh water/oil from (no.) … tank(s) to (no.) … tank(s) to improve the stability.

.7.2 Fore peak/after peak tank was filled/emptied to change the trim.

.8 Cargo was restowed in (no.) … hold(s)/on deck to improve the stability.

선장/기관장이 선교/기관(제어)실에 호출되어 당직사관으로부터 당직을 공식적으로 인수할 경우는, 확인하고 다음과 같이 말해야 함:

.2 나는 지금 당직을 인수함.

당직 사관은 확인을 하고, 다음과 같이 말하여야 함:

.2.1 선장/기관장님이 지금 당직을 인수하셨음.

B1/2 트림, 횡경사 및 복원성

.1 (현재) 본선은 등흘수임.
.1.1 본선은 (현재)선수/선미가 …미터 더 가라앉았음.
.2 (현재)횡경사는 없음.
.2.1 (현재) 좌현/ 우현 측으로 …도 횡경사 되어 있음.
.3 횡경사를 바로잡기 위하여 연료/ 밸러스트/ 청수/ 유류가 (번호)… 탱크로부터 (번호)… 탱크로 이송되었음.
.3.1 횡경사를 바로잡기 위하여 연료/ 밸러스트/ 청수/ 유류를 (번호)… 탱크로부터 (번호)…탱크로 이송하여야함.
.4 횡경사를 바로잡기 위하여 갑판상의/(제 …번 화물창의) 화물이 재적부 되었음.
.4.1 횡경사를 바로잡기 위하여 (…) 창에 화물을 재적부해야 함.
.5 (현재의) 복원성은 양호함/불량함.
.6 복원성을 좋게 하기 위해 (…) 이중저 탱크가 채워졌음.
.7 복원성을 향상시키기 위하여 연료/ 밸러스트/ 청수/ 유류가 (번호) …탱크에서 (번호)…탱크로 이송되었음.
.7.1 복원성을 향상시키기 위하여 연료/ 밸러스트/ 청수/ 유류를 (번호) …탱크에서 (번호)…탱크로 이송하여야함.
.7.2 트림을 바꾸기 위해 선수/선미탱크가 채워졌음/비워졌음.
.8 복원성을 향상시키기 위하여 (…) 화물창/갑판의 화물이 재적부 되었음.

.8.1 We must restow cargo in (no.) ··· hold(s)/on deck to improve the stability.
.9 Containers were restowed from ··· to ··· to improve the stability.
.9.1 We must restow containers from ··· to ··· to improve the stability.

B2 Safety on board

B2/1 General activities

The phrases of this section apply to most of the emergencies covered in this chapter.

B2/1.1 Raising alarm

.1 Operate the general emergency alarm.
.2 Inform the Master/Chief Engineer/··· .
.3 Inform the ··· coast radio station/vessels in vicinity (on radio).
.4 Request assistance (on radio) from ··· and report.
.4.1 Assistance was
~ requested from ··· .
~ offered by ··· .
~ accepted from ··· .
.5 Transmit a SECURITE/PAN-PAN/distress alert/MAYDAY and report.
.5.1 A SECURITE/PAN-PAN/distress alert/MAYDAY was transmitted.
.6 Was the distress alert/MAYDAY acknowledged?
.6.1 Yes, the distress alert/MAYDAY was acknowledged by ···coast radio station/MRCC/vessel(s) in vicinity.
.6.2 No, the distress alert was not acknowledged (yet).
.6.3 Repeat the distress alert.

.8.1 복원성을 향상시키기 위하여 (번호)··· 화물창/ 갑판의 화물 을 재적부 하여야함.

.9 복원성을 향상시키기 위하여 컨테이너가 ···에서 ···으로 재적부 되었음.

.9.1 복원성을 향상시키기 위하여 컨테이너를 ··· 에서 ··· 으로 재적부하여야 함.

B2 선내 안전

B2/1 일반 사항

이 절의 어귀는 이 장 내의 대부분의 비상 상항에 적용됨.

B2/1.1 경보의 발령

.1 일반 비상경보를 작동하시오.

.2 선장/기관장/···에게 알리시오.

.3 ···연안 무선국/근처의 선박에게 (무선으로)알리시오.

.4 (무선으로)···로부터의 원조를 요청하고 보고하시오.

.4.1 ···로부터의 원조를 요청하였음.
···로부터의 원조 제의가 있었음.
···로부터의 원조의 수락이 있었음.

.5 시큐리티/팡 팡/조난경보/메이데이를 발신하고 보고하시오.

.5.1 시큐리티/팡 팡/조난경보/메이데이를 발신하였음.

.6 조난 경보/메이데이는 수신 확인되었는가?

.6.1 예, 조난 경보/메이데이는 ···연안 무선국/해상 구조조정본부/근처 선박에 의하여 수신 확인되었음.

.6.2 아니오, 조난 경보는 (아직) 수신 확인되지 않았음.

.6.3 조난 경보를 반복하시오.

B2/1.2 Briefing crew and passengers

Also see B4 "Passenger Care"

.1 Make the following announcement (on the PA-system):
.2 This is your captain speaking.
.2.1 We have grounded/a minor flooding/a minor fire in ….
.2.2 There is no immediate danger to crew, passengers or vessel - and there is no reason to be alarmed.
.2.3 For safety reasons I request all crew members to go to their assembly stations.
.2.3.1 All officers to report to the bridge.
.2.3.2 Watchkeepers remain at stations until further order.
.2.4 As soon as I have further information I will make another announcement - there is no danger at this time.
.2.5 Fire fighting teams/damage control teams are fighting the fire/flooding.
.2.6 We also have radio contact with other vessels/coast radio stations.
.2.7 The fire/flooding is under control.
.3 This is your captain speaking. I have another announcement:
.3.1 The fire/flooding is not under control yet.
.3.2 Leave the engine room/superstructure/your stations/ your cabins/ … immediately - close all openings.
.3.3 Take lifejackets with you.
.3.3.1 Take your emergency equipment with you according to the muster list.
.3.4 Stand by fire fighting stations/damage control stations and report.
.3.4.1 Fire fighting stations/damage control stations are standing by.
.3.5 All crew members to assembly stations.
.3.6 Follow the escape routes shown.
.3.7 The route to the assembly station … is not clear.

B2/1.2 승무원과 여객에 대한 상황 설명

(B4의 "여객 관리" 편도 참조)

.1 (확성 방송 장치로) 다음의 공고를 하시오:

.2 본선의 선장이 말하겠음.

.2.1 본선은 좌초/약간의 침수/…에 약간의 화재가 발생하였음.

.2.2 승무원, 여객, 또는 선박에 당면한 위험은 없음
- 따라서 경보를 울려야 할 이유는 없음.

.2.3 안전상의 이유로 본인은 모든 승무원으로 하여금 각자의 집합 장소로 가기를 요청함.

.2.3.1 모든 사관들은 선교에 출두하기 바람.

.2.3.2 당직 근무자들은 추가의 지시가 있을 때까지 당직부서에 남기 바람.

.2.4 추가의 통보사항이 있으면 즉시 본인은 추가 방송을 하겠음
- 현재로서는 위험이 없음.

.2.5 소화반/손상 제어반은 소화작업중/방수작업중임.

.2.6 본선은 타 선박들/연안 무선국들과 무선 교신을 하고있음.

.2.7 화재/침수는 진압되었음.

.3 본선의 선장이 말하겠음. 추가의 공고를 하겠음:

.3.1 화재/침수는 아직 진압되지 않았음.

.3.2 기관실/상부 구조물/자기 부서/자기 선실/…에서 즉시 떠나시오
- 모든 개구를 폐쇄하시오.

.3.3 각자 구명동의를 지참하시오.

.3.3.1 부서 배치표에 따라서 자기의 비상 용구를 지참하시오.

.3.4 소화 부서/손상 제어 부서는 대기하고 보고하시오.

.3.4.1 소화 부서/손상 제어 부서는 대기하고 있음.

.3.5 모든 승무원은 집합 장소로 가시오.

.3.6 안내된 탈출 통로를 따라가시오.

.3.7 집합 장소 …로 가는 통로는 명확하지 않음.

.3.7.1 The route to the assembly station will be via … .
.3.8 Assemble
~ on deck.
~ on the foredeck/afterdeck.
~ on the … deck on port side/starboard side.
~ on the … deck forward of …/aft of … .
.3.9 Do not
~ go to the lifeboat/liferaft stations before ordered.
~ enter the lifeboats/liferafts - the order to enter will be given from the bridge/by the officers.
.3.10 The following department(s)/crew members will (temporarily) disembark for safety reasons.

B2/1.3 Checking status of escape routes

.1 Check the escape routes and report.
.1.1 All escape routes are clear.
.1.2 The escape route(s) from … (to …)/via … is/are blocked /not clear (yet).
.1.3 The escape route(s) from … (to …)/via … will be clear in … minutes.

B2/1.4 Checking status of lifeboats/liferafts

.1 Check the launching tracks and report.
.1.1 All launching tracks are clear.
.1.2 The launching track(s) of no. … lifeboat/liferaft is/are not clear (yet).
.1.3 The launching track(s) of no. … lifeboat/liferaft will be clear in … minutes.
.2 Check the working parts and report.
.2.1 All working parts are free.

.3.7.1 집합 장소로 가는 통로는 …를 경유하게 됨.
.3.8 갑판에 집합하시오.
전부갑판/후부갑판에 집합하시오.
좌현/우현 측의 …갑판에 집합하시오.
…의 전부/…의 후부에 있는 …갑판에 집합하시오.
.3.9 지시되기 전에는 구명정/구명벌 부서에 가지 마시오.
구명정/구명벌에 들어가지 마시오-들어가라는 명령은 선교로 부터/사관에 의하여 내려질 것임.
.3.10 다음의 부서/승무원은 안전상의 이유로 (임시) 하선하겠음.

B2/1.3 탈출통로 상태의 점검

.1 탈출통로를 점검하고 보고하시오.
.1.1 모든 탈출통로는 깨끗함(장애물이 없음).
.1.2 …에서(…까지)의/…를 경유하는 탈출통로는 막혔음./(아직) 깨끗하지 않음(장애가 있음).
.1.3 …에서(…까지)의/…를 경유하는 탈출통로는 …분 후 장애가 없어짐.

B2/1.4 구명정/구명벌 상태의 점검

.1 진수용 궤도를 점검하고 보고하시오.
.1.1 모든 진수용 궤도는 깨끗함(장애가 없음).
.1.2 제 …번 구명정/구명벌의 진수용 궤도는(아직) 장애가 있음.
.1.3 제 …번 구명정/구명벌의 진수용 궤도는 …분 후 장애가 없어짐.
.2 작동부를 조사하고 보고하시오.
.2.1 모든 작동부는 자유스러움.

.2.2 The roll(s)/block(s)/rigging/… of no. … lifeboat is/are not free (yet).
.2.3 The roll(s)/block(s)/rigging/… of no. … lifeboat will be free in … minutes.
.3 Check the securings of the launching appliances and report.
.3.1 All securings are in the correct position.
.3.2 The securing of no. … lifeboat/liferaft is not in the correct position.
.3.2.1 Correct the position of the securing.
.3.3 The securing of no. … lifeboat/liferaft is damaged.
.3.3.1 Replace/repair the securing.
.3.4 The harbour pin(s) of no. … lifeboat is/are missing.[98)]
.3.4.1 Replace the harbour pin(s).
.4 Check the fuel/oil of the lifeboat engine(s) and report.
.4.1 The fuel tank of no. … lifeboat engine is full/not full.
.4.1.1 Fill up fuel.
.4.2 The oil level of no. … lifeboat engine is normal/below normal.
.4.2.1 Fill up oil.
.5 Operate the lifeboat engine(s) and report.
.5.1 All lifeboat engines are operational.
.5.2 No. … lifeboat engine is not operational (yet).
.5.3 No. … lifeboat engine will be operational in … minutes.
.6 Check the bilge pumps of the lifeboats and report.
.6.1 All bilge pumps are operational.
.6.2 The bilge pumps of no. … lifeboat are not operational (yet).
.6.3 The bilge pumps of no. … lifeboat will be operational in … minutes.
.7 Check the drain plugs and report.
.7.1 All drain plugs are available.
.7.2 The drain plug(s) in no. … lifeboat is/are missing.
.7.2.1 Replace the drain plug(s).
.8 Check the slip gear in the lifeboats and report.[99)]

.2.2 제 …번 구명정의 바퀴/ 도르레/ 삭구장치/ …이 (아직) 자유스럽지 못함.

.2.3 제…번 구명정의 바퀴/ 도르레/ 삭구장치/ …은 …분 후 자유스러워짐.

.3 진수설비의 고박장치를 점검하고 보고하시오.

.3.1 모든 고박장치는 올바른 위치임.

.3.2 제 …번 구명정/구명벌의 고박장치는 올바른 위치가 아님.

.3.2.1 고박장치의 위치를 바로 잡으시오.

.3.3 제 …번 구명정/구명벌의 고박장치는 손상되었음.

.3.3.1 고박장치를 교환/수리하시오.

.3.4 제 …번 구명정의 하아버 핀이 없어졌음.

.3.4.1 하아버 핀을 새로 끼우시오.

.4 구명정 엔진의 연료/유류를 점검하고 보고하시오.

.4.1 제…번의 구명정 엔진의 연료 탱크가 가득 찼음/가득 차지 않았음.

.4.1.1 연료를 채우시오.

.4.2 제…번의 구명정 엔진의 유류 레벨이 정상임/정상 이하임.

.4.2.1 유류를 채우시오.

.5 구명정 엔진을 작동하고 보고하시오.

.5.1 모든 구명정 엔진이 작동 가능함.

.5.2 제…번의 구명정 엔진이 (아직) 작동되지 않음.

.5.3 제…번의 구명정 엔진이 …분 후에 작동 가능함.

.6 구명정의 빌지 펌프를 점검하고 보고하시오.

.6.1 모든 빌지 펌프는 작동 가능함.

.6.2 제 …번 구명정의 빌지 펌프는(아직) 작동되지 않음.

.6.3 제 …번 구명정의 빌지 펌프는 …분 후 작동 가능함.

.7 배수 플러그를 점검하고 보고하시오.

.7.1 모든 배수 플러그는 사용 가능함.

.7.2 제 …번 구명정의 배수 플러그는 없어졌음.

.7.2.1 배수 플러그를 제자리에 두시오.

.8 구명정의 슬립 기어를 점검하고 보고하시오.

.8.1 All slip gear is in the correct position and secured.
.8.2 The slip gear of no. ··· lifeboat is not in the correct position.
.8.2.1 Correct the position of the slip gear.
.8.3 The slip gear of no. ··· lifeboat is not secured.
.8.3.1 Secure the slip gear.
.9 Check the lifeboat equipment and report.
.9.1 All lifeboat equipment is complete and operational.
.9.2 The lifeboat equipment is not complete.
.9.2.1 Complete the lifeboat equipment.
.10 Launch/hoist no. ··· lifeboat(s) and report.
.10.1 The launching appliances are operational.
.10.2 The launching appliances are not operational.
.10.3 No. ··· winch/davit is not operational (yet).
.10.3.1 No. ··· winch/davit will be operational in ··· minutes.
.10.4 Hoist no. ··· lifeboat(s).
.11 Secure the lifeboat(s) and report.
.11.1 Lifeboat(s) is/are secured.
.12 Check the liferafts and report.
.12.1 All liferafts are in position and operational.
.12.2 No. ··· liferaft(s) is/are not operational.
.12.3 The inflation cord of no. ··· liferaft is not secured on board.[100)]
.12.3.1 Secure the inflation cord.
.12.4 No. ··· liferaft container is damaged.
.12.4.1 Replace the liferaft container in the next port.
.12.5 The inspection tag of no. ··· liferaft is expired.
.12.5.1 Replace the liferaft in the next port.

B2/1.5 Ordering evacuation

.1 Evacuate all rooms/spaces/decks/··· and report.
.1.1 All rooms/spaces/decks/··· evacuated.

.8.1 모든 슬립 기어는 올바른 위치에서 안전하게 되어있음.
.8.2 제 …번 구명정의 슬립 기어는 올바른 위치에 있지 않음.
.8.2.1 슬립 기어의 위치를 바로잡으시오.
.8.3 제…번 구명정의 슬립 기어는 안전하게 되어있지 않음.
.8.3.1 슬립 기어를 안전하게 하시오.
.9 구명정 장비를 점검하고 보고하시오.
.9.1 모든 구명정 장비는 완비되었고 작동 가능함.
.9.2 구명정 장비는 완비되어 있지 않음.
.9.2.1 구명정 장비를 완비하시오.
.10 제 …번 구명정을 진수/인양하고 보고하시오.
.10.1 진수 장치가 작동 가능함.
.10.2 진수 장치가 작동되지 않음.
.10.3 제…번 윈치/대빗트가(아직) 작동되지 않음.
.10.3.1 제…번 윈치/대빗트가…분 후 작동 가능함.
.10.4 제 …번 구명정을 들어올리시오.
.11 구명정을 고박 하고 보고하시오.
.11.1 구명정이 고박 되었음.
.12 구명벌을 점검하고 보고하시오.
.12.1 모든 구명벌들이 제자리에 있고 작동 가능함.
.12.2 제 …번 구명벌이 작동되지 않음.
.12.3 제 …번 구명벌의 팽창줄이 선상에 고박되어 있지 않음.
.12.3.1 팽창줄을 고박하시오.
.12.4 제 …번 구명벌의 용기가 손상되었음.
.12.4.1 다음 항구에서 구명벌의 용기를 교체하시오.
.12.5 제 …번 구명벌의 검사 표지의 유효기간이 경과되었음.
.12.5.1 다음 항구에서 구명벌을 교체하시오.

B2/1.5 대피 명령

.1 모든 선실/장소/갑판/…에서 대피하고 보고하시오.
.1.1 모든 선실/장소/갑판/…에서 대피하였음.

.2 Evacuate engine room and report.
.2.1 Engine room evacuated.
.3 Evacuate no. ··· hold(s)/tank(s) and report.
.3.1 No. ··· hold(s)/tank(s) evacuated.
.4 Evacuate superstructure and report.
.4.1 Superstructure evacuated.
.5 Evacuate accommodation and report.
.5.1 Accommodation evacuated.
.6 Do not enter ··· deck/space/area.
.7 Report missing persons/injured persons/casualties.
.7.1 No persons missing/injured.
.7.2 Number of missing persons/injured persons/casualties is: ···.
.7.3 ··· deck/space/area not accessible (yet).
.8 Provide first aid (in the vessel's hospital/at a safe place).
.8.1 Request medical assistance from ··· (on radio).
.9 All persons are outside the danger area.

B2/1.6 Roll call

.1 Report number of all persons/passengers/crew members at assembly stations.
.1.1 Number of all persons/passengers/crew members at assembly station ··· is ··· .
.1.2 Number of persons/passengers/crew members at assembly station ··· is complete.
.1.3 Number of persons/passengers/crew members at assembly station ··· is not complete (yet).
.1.4 ··· passenger(s)/crew member(s)is/are missing.
.2 Search for missing passenger(s)/crew member(s) and report.
.2.1 Missing passenger(s)/crew member(s) recovered.
.2.2 Missing passenger(s)/crew member(s) not recovered (yet) -(search is continuing).

.2 기관실에서 대피하고 보고하시오.
.2.1 기관실에서 대피하였음.
.3 제…번 화물창/탱크에서 대피하고 보고하시오.
.3.1 제…번 화물창/탱크에서 대피하였음.
.4 상부 구조물에서 대피하고 보고하시오.
.4.1 상부 구조물에서 대피하였음.
.5 거주 구역에서 대피하고 보고하시오.
.5.1 거주 구역에서 대피하였음.
.6 …갑판/장소/구역에 들어가지 마시오.
.7 실종자/부상자/사상자를 보고하시오.
.7.1 실종자/부상자 없음.
.7.2 실종자/부상자/사상자의 수는 :…명임.
.7.3 …갑판/장소/구역에는 (아직)접근할 수 없음.
.8 (선내의 의무실에서/ 안전한 장소에서)응급처치를 제공 하시오.
.8.1 (무선으로) …에 의료 원조를 요청하시오.
.9 모든 사람은 위험 구역의 밖에 있음.

B2/1.6 인원 점검

.1 집합 장소에 있는 모든 사람/ 여객/ 승무원의 수를 보고하시오.
.1.1 집합 장소…에 있는 모든 사람/ 여객/ 승무원의 수는 …명임.
.1.2 집합 장소…에 있는 모든 사람/ 여객/승무원의 수는 완전함.
.1.3 집합 장소…에 있는 사람/ 여객/ 승무원의 수는 (아직) 완전치 않음.
.1.4 여객/승무원 …명이 실종됨.
.2 실종된 여객/ 승무원을 수색하고 보고하시오.
.2.1 실종된 여객/ 승무원은 되찾았음.
.2.2 실종된 여객/ 승무원은 (아직) 되찾지 못했음-(수색이 계속됨)

.3 Watchkeepers to assembly stations.
.4 Lifeboatmen! Check the equipment of the crew at assembly stations and report.
.4.1 Equipment of the crew at assembly station … is complete.
.4.2 Equipment of the crew at assembly station …is not complete(yet).
.4.3 Complete the equipment and report.
.4.3.1 Go for blanket/stretcher/… and report.
.5 Lifeboatmen! Check the outfit of the passengers at assembly stations and report.
.5.1 Outfit of the passengers at assembly station … is correct.
.5.2 Outfit of the passengers at assembly station … is not correct (yet).
.5.2.1 Correct the outfit and report.
.5.2.2 Put on warm clothing/long sleeved shirt/long trousers/strong shoes/head covering/… and report.
.6 Passengers and crew! Follow the lifeboatmen to the lifeboat stations/liferaft stations on the embarkation deck.

B2/1.7 Ordering abandon vessel

.1 Swing out no. … lifeboat(s) and report.
.1.1 No. … lifeboat(s) swung out.
.2 Lower no. … lifeboat(s) alongside the embarkation deck and report.
.2.1 No. … lifeboat(s) is/are alongside the embarkation deck.
.3 Enter the lifeboat(s) (no. …) and report.
.3.1 Enter the lifeboat(s)/liferaft(s) over the … deck.
.3.2 Enter the lifeboat(s)/liferaft(s) over the ladders/nets/manropes.
.3.3 Jump into the water and enter the lifeboat(s)/liferaft(s).
.3.4 Jump onto the liferaft(s) alongside the vessel.
.3.5 Do not push each other when entering.
.3.6 Assist injured/helpless persons.
.3.7 Clear the entrance of the lifeboat/liferaft.

.3 당직자들은 집합 장소로 오시오.
.4 구명정수! 집합 장소의 승무원의 장비들을 점검하고 보고하시오.
.4.1 집합 장소…의 승무원의 장비는 완전함.
.4.2 집합 장소…의 승무원의 장비는 (아직) 완전치 못함.
.4.3 장비를 완전하게하고 보고하시오.
.4.3.1 담요/들것/…을 가져오고 보고하시오.
.5 구명정수! 집합 장소의 여객의 장구(裝具)를 점검하고 보고하시오.
.5.1 집합 장소…의 여객의 장구는 올바름.
.5.2 집합 장소…의 여객의 장구는 (아직) 올바르지 못함.
.5.2.1 장구를 바로잡고 보고하시오.
.5.2.2 보온복/소매 긴 셔츠/긴 바지/튼튼한 신발/머리 보호구/…을 착용하고 보고하시오.
.6 여객 및 승무원들! 구명정수를 따라서 승정갑판에 있는 구명정 부서/구명벌 부서로 가시오.

B2/1.7 퇴선 명령

.1 제 …번 구명정을 선외로 내고 보고하시오.
.1.1 제 …번 구명정을 선외로 내었음.
.2 제 …번 구명정을 승정갑판에 평행하게 내리고 보고하시오.
.2.1 제 …번 구명정이 승정갑판에 나란히 되었음.
.3 (…번) 구명정에 들어가서 보고하시오.
.3.1 …갑판을 통하여 구명정/구명벌에 들어가시오.
.3.2 사다리/그물/손잡이 줄을 통해 구명정/구명벌에 들어가시오.
.3.3 물속으로 뛰어든 다음 구명정/구명벌에 탑승하시오.
.3.4 본선에 접안한 구명벌 위에 뛰어내리시오.
.3.5 들어갈 때 서로 밀지 마시오.
.3.6 부상자/거동이 불편한 자를 도우시오.
.3.7 구명정/구명벌의 입구에 있지 마시오.

.3.8 Sit down in the lifeboat/liferaft immediately.
.3.9 Hold on to the ropes or to your seat when launching.
.4 No. … lifeboat(s)/liferaft(s) entered.
.5 Let go no. … lifeboat(s)/liferaft(s) and report.
.5.1 No. … lifeboat(s)/liferaft(s) is/are let go.
.6 Throw over board no. … liferaft and report.
.6.1 No. … liferaft thrown over board.
.7 Inform coast radio stations/vessels in vicinity about the number of lifeboats/liferafts launched and report.
.7.1 Inform coast radio stations/vessels in vicinity about the number of persons in each lifeboat/liferaft and report.
.7.2 Inform coast radio stations/vessels in vicinity about the number of crew members remaining on board.
.7.3 Coast radio station …/vessels in vicinity informed.
.8 Stand clear of the vessel and report.
.8.1 No. … lifeboat(s)/liferaft(s) standing clear.
.8.2 No. … lifeboat(s)/liferaft(s) not standing clear.
.9 Rescue boat/no. … motor lifeboat!
Assist no. … lifeboat(s)/liferaft(s) and report.
.9.1 Rescue boat/no. … motor lifeboat is assisting.
.9.2 No. … lifeboat(s)/liferaft(s) standing clear of the vessel now.

B2/1.8 In-boat procedures

Also see: B4 "Passenger Care" 2.5 and 2.6.

.1 Stand by engine/pumps/lookout/entrance and report.
.1.1 Engine/pumps/lookout/entrance is/are standing by.
.2 Recover persons in water and report.
.2.1 Number of persons recovered is: … .
.2.2 Keep lookout for further persons in water.
.2.3 Report the total number of persons in lifeboat(s)/liferaft(s).
.2.3.1 The total number of persons is now: … .

.3.8 구명정/구명벌 내에서는 즉시 앉으시오.
.3.9 진수시에 밧줄이나 각자의 좌석을 붙잡으시오.
.4 제…번 구명정/구명벌에 탑승하였음.
.5 제…번 구명정/구명벌을 풀어주고 보고하시오.
.5.1 제…번 구명정/구명벌이 풀어주었음.
.6 제…번 구명벌을 선외로 투하하고 보고하시오.
.6.1 제…번 구명벌이 선외로 투하되었음.
.7 진수된 구명정/ 구명벌의 숫자를 연안 무선국/근처의 선박에 게 통보하고 보고하시오.
.7.1 각 구명정/구명벌 내의 사람 수를 연안 무선국/ 근처의 선박에 게 통보하고 보고하시오.
.7.2 선내에 남아있는 승무원의 수를 연안 무선국/ 근처의 선박에 게 통보하시오.
.7.3 …연안 무선국/근처의 선박에게 통보되었음.
.8 본선으로부터 벗어나서 보고하시오.
.8.1 제…번 구명정/구명벌은 선박에서 벗어나 있음.
.8.2 제…번 구명정/구명벌은 벗어나지 못했음.
.9 구조정/제…번 모터 구명정!
제…번 구명정/구명벌을 원조하고 보고하시오.
.9.1 구조정/제…번 모터 구명정은 원조하고 있음.
.9.2 제…번 구명정/구명벌은 현재 본선으로부터 벗어나 있음.

B2/1.8 구명정 내의 절차

(B4 의 "여객 관리" 2.5 와 2.6도 참조할 것)

.1 기관/펌프/견시/입구를 준비하고 보고하시오.
.1.1 기관/펌프/견시/입구는 준비하고 있음.
.2 물속에 있는 사람을 끌어올리고 보고하시오.
.2.1 끌어올린 사람의 수는: …명임.
.2.2 물속에 사람이 더 있는지 견시를 유지하시오.
.2.3 구명정/구명벌 내의 전 인원수를 보고하시오.
.2.3.1 현재의 전 인원수는: …명임.

.3 Report the number of injured persons.
.3.1 No persons injured.
.3.2 The number of injured persons is: … .
.3.3 Provide first aid to injured persons.
.3.4 Secure injured/helpless persons.
.4 Let go sea anchor and report.
.4.1 Sea anchor is let go.
.5 Report the number of lifeboats/liferafts in sight.
.5.1 The number of lifeboats/liferafts in sight is: … .
.6 Contact the lifeboat(s)/liferaft(s) on radio and report.
.6.1 Lifeboat(s)/liferaft(s) contacted.
.6.2 No contact possible.
.7 Give distress signals for identification.
.7.1 Fire rockets for identification.
.7.2 Use glasses/lamps/mirrors for identification.
.7.3 Give sound signals/… signals for identification.
.8 Start the engine and report.
.9 Set sail.
.10 Use oars.
.11 Join the other lifeboat(s)/liferaft(s).
.11.1 Connect the lifeboats/liferafts with lines and report.
.11.2 … lifeboats/liferafts connected.

B2/2 Occupational safety

B2/2.1 Instruction

.1 Prepare a training plan for occupational safety.
.2 When was the last training session on occupational safety?
.2.1 The last training session was on … *(date).*
.3 When is the next training session on occupational safety?
.3.1 The next training session is on … *(date).*

.3　부상자의 수를 보고하시오.
.3.1　부상자는 없음.
.3.2　부상자의 수는: …명임.
.3.3　부상자에게 응급처치를 제공하시오.
.3.4　부상자/거동이 불편한 자를 안전하게 하시오.
.4　해묘를 투묘하고 보고하시오.
.4.1　해묘를 투묘하였음.
.5　시야에 들어오는 구명정/구명벌의 숫자를 보고하시오.
.5.1　시야에 들어오는 구명정/구명벌의 숫자는: …임.
.6　무선으로 구명정/구명벌과 교신하고 보고하시오.
.6.1　구명정/구명벌과 교신되었음.
.6.2　교신이 불가능함.
.7　식별하기 위하여 조난신호를 하시오.
.7.1　식별하기 위하여 로켓을 발사하시오.
.7.2　식별하기 위하여 유리/램프/거울을 사용하시오.
.7.3　식별하기 위하여 음향신호/…신호를 하시오.
.8　엔진을 시동하고 보고하시오.
.9　돛을 설치하시오.
.10　노를 사용하시오.
.11　다른 구명정/구명벌과 합류하시오.
.11.1　구명정/구명벌들을 줄로 연결하고 보고하시오.
.11.2　…구명정/구명벌들을 연결하였음.

B2/2 직무상의 안전

B2/2.1 교육

.1　직무상의 안전에 대한 훈련 계획을 준비하시오.
.2　직무상의 안전에 대한 지난번의 훈련기간은 언제였는가?
.2.1　지난번의 훈련기간은 …(날짜) 이었음.
.3　직무상의 안전에 대한 다음번의 훈련기간은 언제인가?
.3.1　다음번의 훈련기간은 …(날짜)임.

.4 Are new crew members/passengers instructed on occupational safety?
.4.1 Yes, new crew members/passengers are instructed.
.4.2 No, new crew members/passengers are not instructed (yet).
.4.3 Instruct new crew members/passengers by …*(time)*/on … *(date)*.
.5 Participation in training sessions on occupational safety is mandatory.

B2/2.2 Practical occupational safety

.1 Instruct crew on occupational safety before departure.
.2 Have special instruction on dangerous goods/heavy lifts/ cargo securing/illumination/ventilation/… .
.3 Where are dangerous goods carried on board?
.3.1 Dangerous goods of IMO Class … are carried
~ on deck (in roped-off areas).
~ in no. … hold(s).
~ in … /on… .
.4 Prepare an emergency plan.
.5 Brief all crew members/passengers on the symptoms caused by dangerous substances.
.6 What signals/communications are used in case of emergency?
.6.1 The following signals/communications are used in case of emergency: … .
.7 Brief all crew members/passengers
~ about restricted areas.
~ how to report in/out (when entering/leaving bridge/engine room/…).
.8 Do not enter the unmanned (engine) room/… space without permission.
.8.1 Report on telephone/radio/… while in the (engine) room /…space every … minutes.

.4 신규 승무원/ 여객은 직무상의 안전에 관하여 교육을 받았는가?
.4.1 예, 신규 승무원/여객은 교육을 받았음.
.4.2 아니오, 신규 승무원/여객은 (아직) 교육을 받지 않았음.
.4.3 신규 승무원/여객을 …(시각)까지/…(날짜)에 교육하시오.
.5 직무상의 안전에 대한 훈련에 참가하는 것은 강제적임.

B2/2.2 직무 안전 실무

.1 출항 전에 직무상 안전에 대하여 승무원을 교육하시오.
.2 위험 화물 /중량 화물/화물 고박/조명/통풍/…에 관하여 특별교육을 실시하시오.
.3 선상의 어디에 위험 화물을 싣고 있는가?
.3.1 IMO 등급 …의 위험 화물이 갑판상에 실려 있음.(로프로 격리된 구역에)
IMO 등급 …의 위험 화물이 제…번 화물창에 실려 있음.
IMO 등급 …의 위험 화물이 …안에 /…위에 실려 있음.
.4 비상 계획표를 준비하시오.
.5 위험 물질로 야기되는 증상에 대하여 모든 승무원/여객에게 설명하시오.
.6 비상시에 어떠한 신호/교신이 이용되는가?
.6.1 비상시에 다음의 신호/교신이 이용됨.: … .
.7 모든 승무원/여객에게 제한 구역에 대하여 설명하시오.
모든 승무원/여객에게(선교/기관실/…에 들어갈 때/나갈 때) 입/출보고 방법에 대하여 설명하시오.
.8 허락 없이 무인(기관)실/…장소에 들어가지 마시오.
.8.1 (기관)실/…장소에 있을 때 …분마다 전화/무선/…으로 보고하시오.

.9 Brief all crew members/passengers on the storm.

.9.1 **Attention!**
Entering the forecastle/main deck/weather side/… of the vessel is prohibited/dangerous (due to storm).

.9.2 **Attention!**
Make use of handrails and lifelines in corridors and on deck.

.9.3 **Attention!**
Close all dead lights and storm doors.

.9.4 Attention!
Secure all loose objects in your cabins/on deck/in … .

.10 Brief all crew members/passengers on winter conditions /tropical conditions.

.11 Check the completeness and availability of the occupational safety equipment and report.

.11.1 Occupational safety equipment is complete and available.

.11.2 Following occupational safety equipment is not complete/ available: … .

.11.3 Occupational safety equipment will be complete and available in … hour(s).

.12 Appoint an officer/a crew member in charge of safety before working.

.13 Take additional safety measures for the
~ work on masts.
~ work outboard.
~ work in hold(s)/tank(s).
~ work in extreme weather conditions/… .

B2/2.3 Occupational accidents

.1 Accident in engine room/in no. … hold/in no. … tank/in superstructure/in accommodtaion/in … space/on deck/ outboard/on pier/on …/in … .

.9 모든 선원/여객에게 폭풍에 대하여 설명하시오.

.9.1 **주의 !**

(폭풍으로 인하여) 선박의 선수/ 주갑판/ 풍상측/ …에 진입하는 것은 금지됨/위험함.

.9.2 **주의 !**

통로와 갑판 상에서는 난간과 구명줄을 활용하시오.

.9.3 **주의 !**

모든 현창 뚜껑과 방풍 문을 폐쇄하시오.

.9.4 **주의 !**

각자의 선실/ 갑판/ …에 있는 모든 느슨한 물체들을 고박하시오.

.10 동계/열대 상황에 대하여 모든 승무원/여객에게 설명하시오.

.11 직무상의 안전장비에 대한 완전성과 사용 가능성을 점검 하고 보고 하시오.

.11.1 직무상의 안전장비는 완전하고 사용 가능함.

.11.2 다음의 직무상의 안전장비는 완전/사용 가능하지 않음: … .

.11.3 직무상의 안전장비는 …시간 후에 완전하고 사용 가능함.

.12 작업 전에 사관/선원 한 명을 안전 책임자로 임명하시오.

.13 마스트에서 작업하기 위한 추가의 안전조치를 취하시오.

선외에서 작업하기 위한 추가의 안전조치를 취하시오.

화물창/탱크에서 작업하기 위한 추가의 안전조치를 취하시오.

극한 기상상태/ …에서 작업하기 위한 추가의 안전조치를 취하시오.

B2/2.3 직무상의 사고

.1 기관실/ 제 …번 화물창/ 제…번 탱크/ 상부 구조물/ 거주구역/ …장소/ 갑판상/ 선외/ 부두/ …위/ …안에서 사고가 발생함.

.2 Report injured persons/casualties.
.2.1 No person injured.
.2.2 The number of injured persons/casualties is: … .
.3 What happened?
.3.1 Explosion/fire in … .
.3.2 Accident with cargo.
.3.3 Fall from …/into … .
.3.4 Electrical accident in … .
.3.5 Leakage of gas/… .
.3.6 … .
.4 Take immediate action to recover injured person(s) /casualties.
.4.1 Provide first aid.
.4.2 Take immediate action to control the danger area.
.5 What kind of assistance is required?
.5.1 No assistance is required.
.5.2 Medical/technical assistance is required.
.5.3 Shoreside assistance is required.
.6 Secure the danger area and report.
.6.1 The danger area is secured.
.7 Prepare an accident report.

B2/3 Fire protection and fire fighting

B2/3.1 Fire protection

.1 Checking status of equipment

.1 Have fire patrols (every … hour(s)/… time(s) every watch).
.1.1 Have fire patrols
~ in all spaces.
~ in the engine room/cargo hold(s)/superstructures /accommodation/ … .
~ on deck.

.2 부상자/사상자를 보고하시오.
.2.1 부상자가 없음.
.2.2 부상자/사상자의 숫자는: …명임.
.3 무슨 일이 일어났는가?
.3.1 …에서의 폭발/화재임.
.3.2 화물에 관련된 사고임.
.3.3 …에서/…으로 추락하였음.
.3.4 …에서의 전기 사고임.
.3.5 가스/…의 누출임.
.3.6 …임.
.4 부상자/사상자를 회복시키기 위하여 즉시 조치를 취하시오.
.4.1 응급처치를 제공하시오.
.4.2 위험 구역을 제어하기 위하여 즉각적인 조치를 취하시오.
.5 어떤 종류의 원조가 필요한가?
.5.1 원조는 필요 없음.
.5.2 의료/기술상의 원조가 필요함.
.5.3 육상의 원조가 필요함.
.6 위험 구역을 안전하게 하고 보고하시오.
.6.1 위험 구역은 안전하게 되었음.
.7 사고 보고서를 준비하시오.

B2/3 화재 예방과 소화 작업

B2/3.1 화재 예방

.1 설비 상태의 점검

.1 (매 …시간마다/매 당직마다…번씩)화재 순찰을 하시오.
.1.1 모든 장소에서 화재 순찰을 하시오.
기관실/ 화물창/ 상부 구조물/ 거주 구역/…에서 화재 순찰을 하시오.
갑판 상에서 화재 순찰을 하시오.

.1.2 Have a permanent fire watch.
.2 Is everything in order?
.2.1 Yes, everything is in order.
.2.2 No, following is not in order: … .
.3 Check the fire/smoke alarm(s) and report.
.3.1 All fire/smoke alarms are operational.
.3.2 Fire/smoke alarm(s) in … is/are not operational (yet).
.3.3 Fire/smoke alarm(s) in … will be operational in … minutes.
.4 Check the portable extinguishers and report.
.4.1 All portable extinguishers are in position and operational.
.4.2 The portable extinguishers in …
~ are not in position (yet).
~ will be in position in … minutes.
~ are not accessible (yet).
~ will be accessible in … minutes.
~ are missing.
.4.2.1 Replace the missing portable extinguisher(s).
.4.3 The inspection tag(s) of the portable extinguisher(s) in … is/ are broken/expired.
.4.3.1 Replace the portable extinguisher(s) with broken/expired inspection tag(s).
.5 Check the fire mains and report.
.5.1 All fire mains are operational.
.5.2 The hydrant(s) in … is/are not operational (yet).
.5.2.1 The hydrant(s) will be operational in … minutes.
.5.3 The hose(s) to hydrant(s) in … is/are worn/cut.
.5.3.1 Replace the worn/cut hose(s).
.5.4 The hose(s)/spanner(s)/nozzle(s) to hydrant(s) in … is/are missing.
.5.4.1 Replace the missing hose(s)/spanner(s)/nozzles(s).
.5.5 The fire pump(s) in … is/are not operational (yet).

.1.2 영구적인 화재 당직을 유지하시오.
.2 모든 것이 정상인가?
.2.1 예, 모든 것이 정상임.
.2.2 아니오, 다음의 사항은 정상이 아님:… .
.3 화재/연기 경보기를 점검하고 보고하시오.
.3.1 모든 화재/연기 경보기는 작동 가능함.
.3.2 …안의 화재/연기 경보기는 (아직)작동이 안됨.
.3.3 …안의 화재/연기 경보기는 …분 후에 작동 가능함.
.4 휴대용 소화기를 점검하고 보고하시오.
.4.1 모든 휴대용 소화기는 제자리에 있고 작동 가능함.
.4.2 …내에 있는 휴대용 소화기는 (아직) 제자리에 없음.
…내에 있는 휴대용 소화기는 …분 후 제자리에 있을 것임.
…내에 있는 휴대용 소화기에는 (아직) 접근할 수 없음.
…내에 있는 휴대용 소화기에는 …분 후에 접근할 수 있음.
…내에 있는 휴대용 소화기는 없어졌음.
.4.2.1 없어진 휴대용 소화기를 다시 비치하시오.
.4.3 …내에 있는 휴대용 소화기의 검사표지가 망가졌음/ 유효 기간이 지났음.
.4.3.1 검사 표지가 망가진/유효기간이 지난 휴대용 소화기를 교체 하시오.
.5 소화 주관을 점검하고 보고하시오.
.5.1 모든 소화 주관은 작동 가능함.
.5.2 …내에 있는 소화전은 (아직)작동되지 않음.
.5.2.1 소화전은 …분 후 작동 가능함.
.5.3 …내에 있는 소화전의 호스는 마멸/절단되었음.
.5.3.1 마멸/절단된 호스를 교체하시오.
.5.4 …내에 있는 소화전의 호스/스패너/노즐은 없어졌음.
.5.4.1 없어진 호스/스패너/노즐을 다시 비치하시오.
.5.5 …내의 소화 펌프는 (아직) 작동되지 않음.

.5.5.1 Fire pump(s) in … will be operational in … minutes.
.5.6 The water pipe(s) in … is/are leaking.
.5.6.1 Repair the leaking water pipe(s) in … .
.5.7 The water pipe(s) in … is/are blocked.
.5.7.1 Free the blocked water pipe(s) in … .
.5.8 Pressure in the water pipe(s) in … is too high/low.
.5.8.1 Reduce/increase pressure in the water pipe(s)in … .
.6 Check the fixed foam/gas fire extinguishing system and report.
.6.1 The fixed foam/gas system is operational.
.6.2 The fixed foam/gas system is not operational (yet).
.6.2.1 The fixed foam/gas system will be operational in … minutes.
.7 Check the sprinkler system and report.
.7.1 The sprinkler system is operational.
.7.2 The sprinkler system in … is not operational (yet).
.7.2.1 The sprinkler system in … will be operational in … minutes.
.8 Check the ventilation system and report.
.8.1 The ventilation system is operational.
.8.2 The ventilation system is not operational (yet).
.8.2.1 The ventilation system will be operational in … minutes.
.8.3 The remote control is not operational (yet).
.8.3.1 The remote control will be operational in … minutes
.8.4 The indicators are not operational (yet).
.8.4.1 The indicators will be operational in … minutes.
.8.5 The fire dampers in … are not operational (yet).
.8.5.1 The fire dampers in … will be operational in … minutes.
.8.6 The fire dampers in … are painted stuck.
.8.6.1 Clear the fire dampers.
.9 Check the skylights/windows/… and report.
.9.1 The skylights/windows/… in/to … are open.
.9.1.1 Close the skylights/windows/… in/to … .
.10 Check the watertight door control and report.

.5.5.1 …내의 소화 펌프는 …분 후에 작동 가능함.
.5.6 …내의 소화수관은 새고 있음.
.5.6.1 …내의 새는 소화수관을 수리하시오.
.5.7 …내의 소화수관은 막혔음.
.5.7.1 …내의 막힌 소화수관을 뚫으시오.
.5.8 …내의 소화수관의 압력은 너무 높음/낮음.
.5.8.1 …내의 소화수관의 압력을 낮추시오/높이시오.
.6 고정식 포말/가스 소화 설비를 점검하고 보고하시오.
.6.1 고정식 포말/가스 소화 설비는 작동 가능함.
.6.2 고정식 포말/가스 소화 설비는 (아직) 작동 되지 않음.
.6.2.1 고정식 포말/가스 소화 설비는 …분 후 작동 가능함.
.7 스프링클러 설비를 점검하고 보고하시오.
.7.1 스프링클러 설비는 작동 가능함.
.7.2 …내의 스프링클러 설비는 (아직) 작동 되지 않음.
.7.2.1 …내의 스프링클러 설비는 …분 후 작동 가능함.
.8 통풍 설비를 점검하고 보고하시오.
.8.1 통풍 설비는 작동 가능함.
.8.2 통풍 설비는 (아직) 작동 되지 않음.
.8.2.1 통풍 설비는 …분 후 작동 가능함.
.8.3 원격 조종 장치는 (아직) 작동 되지 않음.
.8.3.1 원격 조종 장치는 …분 후 작동 가능함.
.8.4 지시기가 (아직)작동 되지 않음.
.8.4.1 지시기는 …분 후 작동 가능함.
.8.5 …내의 소화 댐퍼는 (아직)작동 되지 않음.
.8.5.1 …내의 소화 댐퍼는 …분 후 작동 가능함.
.8.6 …내의 소화 댐퍼는 페인트로 고착됨.
.8.6.1 소화 댐퍼를 자유로이 움직이게 하시오.
.9 천창/창문/…을 점검하고 보고하시오.
.9.1 …내의/…에 있는 천창/창문/…은 열려있음.
.9.1.1 …내의/…에 있는 천창/창문/…을 닫으시오.
.10 수밀문 조정기를 점검하고 보고하시오.

.10.1 The watertight door control is operational.
.10.2 The watertight door control in … is not operational (yet).
.10.3 The watertight door control in … will be operational in … minutes.
.11 Check the electrical lighting and report.
.11.1 The electrical lighting is operational.
.11.2 The electrical lighting in … is not operational (yet).
.11.3 The electrical lighting in … will be operational in … minutes.
.11.4 Switch on/off the electrical lighting in … .
.12 Check the emergency power supply and report.
.12.1 The emergency power supply is operational.
.12.2 The emergency power supply is not operational (yet).
.12.3 The emergency power supply will be operational in … minutes.
.13 Check the firemen's outfits and report.[101)]
.13.1 All firemen's outfits are complete and available.
.13.2 The firemen's outfits are not complete.
.13.2.1 Complete the firemen's outfits.

B2/3.2 Fire fighting and drills

.1 Reporting fire

.1 Fire on board!
.1.1 Smoke/fumes/fire/explosion
~ in engine room.
~ in no. … hold(s)/tank(s).
~ in superstructure/accommodation.
~ in … space.
~ on deck/… .
.1.2 Smoke/fumes from ventilator(s).
.1.3 Burnt smell/fumes in …/ from … .
.2 Report injured persons/casualties.

.10.1　수밀문 조정기는 작동 가능함.
.10.2　…내의 수밀문 조정기는 (아직) 작동 되지 않음.
.10.3　…내의 수밀문 조정기는 …분 후 작동 가능함.
.11　전등을 점검하고 보고하시오.
.11.1　전등은 작동 가능함.
.11.2　…내의 전등은 (아직)작동 되지 않음.
.11.3　…내의 전등은 …분 후 작동 가능함.
.11.4　…내의 전등을 켜시오/끄시오.
.12　비상 전원을 점검하고 보고하시오.
.12.1　비상 전원은 작동 가능함.
.12.2　비상 전원은 (아직)작동 되지 않음.
.12.3　비상 전원은 …분 후 작동 가능함.
.13　소화원 장구를 점검하고 보고하시오.
.13.1　모든 소화원 장구는 완전하고 사용 가능함.
.13.2　소화원 장구는 완전치 않음.
.13.2.1 소화원 장구를 완전하게 하시오.

B2/3.2 소화 작업과 훈련

.1　화재의 보고

.1　선상에 화재 발생!
.1.1　기관실에 연기/ 유해 가스/ 화재/ 폭발이 발생함.
…번 화물창/ 탱크에 연기/ 유해 가스/ 화재/ 폭발이 발생함.
상부 구조물/거주구역에 연기/유해 가스/화재/폭발이 발생함.
…장소에 연기/ 유해 가스/ 화재/ 폭발이 발생함.
갑판 상/…에 연기/ 유해 가스/ 화재/ 폭발이 발생함.
.1.2　통풍기로부터 연기/유해 가스가 나옴.
.1.3　…내에/…로부터 탄 냄새/유해 가스가 나옴.
.2　부상자/사상자를 보고하시오.

.2.1 No person injured.
.2.2 Number of injured persons/casualties is: … .
.3 What is on fire?
.3.1 Fuel/cargo/car(s)/truck(s)/waggon(s)/containers (with dangerous goods)/… on fire.
.3.2 No information (yet).
.4 Is smoke toxic?
.4.1 No, smoke not toxic.
.4.2 Yes, smoke toxic
.5 Is fire under control?
.5.1 Yes, fire (in …) under control.
.5.2 No, fire (in …) not under control (yet).
.5.2.1 Fire spreading (to …).
.5.2.2 Fire (in …) not accessible.
.6 Report damage.
.6.1 No damage.
.6.2 Minor/major damage in …/ to … .
.6.3 No power supply (in …).
.6.4 Making water in … .
.7 Pressure on fire mains!
.8 Shut down main engine(s)/auxiliary engine(s)/… and report.
.8.1 Main engine(s)/auxiliary engine(s)/… shut down.
.9 Stop fuel and report.
.9.1 Fuel stopped.
.10 Close all openings (in …/in all rooms) and report.
.10.1 All openings (in …/in all rooms) closed.
.10.1.1 Openings in … not accessible.
.11 Switch off ventilator(s) (in …) and report.
.11.1 Ventilator(s) (in …) switched off.
.12 Turn bow/stern to windward.
.13 Turn port side/starboard side to windward.
.14 Alter course to … .

.2.1 부상자 없음.
.2.2 부상자/사상자의 숫자는: …명임.
.3 무엇에 불이 났는가?
.3.1 연료/ 화물/ 자동차/ 트럭/ 웨건/(위험 화물 적재된)컨테이너 / …에 불이 났음.
.3.2 (아직)정보가 없음.
.4 연기는 유독성인가?
.4.1 아니오, 연기는 유독성이 아님.
.4.2 예, 연기는 유독성임.
.5 화재는 진압되었는가?
.5.1 예, (…내의) 화재는 진압되었음.
.5.2 아니오, (…내의) 화재는 (아직) 진압되지 않았음.
.5.2.1 화재는 (…로) 확산중임.
.5.2.2 (…내의) 화재에 접근할 수 없음.
.6 손상을 보고하시오.
.6.1 손상 없음.
.6.2 …내에/…에 가벼운/심한 손상이 생김.
.6.3 (…내에) 전원 공급이 안됨.
.6.4 …내에 누수가 생김.
.7 소화 주관에 수압을 가하시오!
.8 주기관/보조기관/…을 정지하고 보고하시오.
.8.1 주기관/보조기관/…이 정지되었음.
.9 연료 공급을 중단하고 보고하시오.
.9.1 연료 공급이 중단되었음.
.10 (…내의/모든 방의)모든 개구를 닫고 보고하시오.
.10.1 (…내의/모든 방의)모든 개구가 닫혔음.
.10.1.1 …내의 개구에는 접근 불능임.
.11 (…내의)통풍기를 끄고 보고하시오.
.11.1 (…내의)통풍기를 껐음.
.12 선수/선미를 풍상측으로 돌리시오.
.13 좌현/우현을 풍상측으로 돌리시오.
.14 …으로 변침 하시오.

.2 Reporting readiness for action

.1 Stand by fire fighting team/rescue team/first aid team/support team and report.

.1.1 Fire fighting team/rescue team/first aid team/support team standing by.

.2 Stand by main engine and report.

.2.1 Main engine standing by.

.3 Stand by CO_2 station/… station/emergency generator.

.3.1 CO_2 station/… station/emergency generator standing by.

.4 Close all openings (in …/in all rooms) and report.

.4.1 All openings (in …/in all rooms) closed.

.4.1.1 Openings in … not accessible.

.3 Orders for fire fighting

.1 Start fire fighting.

.1.1 Take one/two/… fire fighting teams/… team(s) to scene.

.2 Go following route:

.2.1 Go through engine room/no. … hold(s)/tank(s)/superstructure /accommodation/… space/manhole(s) to …space/funnel/….

.2.2 Go from

~ outside/inside to … .

~ port side/starboard side to … .

~ … to … .

.3 Take following (additional) safety measures and report.

.3.1 Have two/… members in one team.

.3.1.1 Number of members in fire fighting team/… team is: … .

.3.2 Have lifeline between each other/to outside.

.3.2.1 … team members have lifelines between each other.

.3.2.2 … team has lifelines to outside.

.2 작업 준비 상황의 보고

.1 소화반/구조반/응급처치반/원조반을 대기하고 보고하시오.

.1.1 소화반/구조반/응급처치반/원조반을 대기하고 있음.

.2 주기관을 대기하고 보고하시오.

.2.1 주기관을 대기하고 있음.

.3 이산화탄소 부서/…부서/비상 발전기를 대기하시오.

.3.1 이산화탄소 부서/…부서/비상 발전기를 대기하고 있음.

.4 (…내의/모든 방의)모든 개구를 닫고 보고하시오.

.4.1 (…내의/모든 방의)모든 개구를 닫았음.

.4.1.1 …의 개구에는 접근 불가능함.

.3 소화 작업 관련 명령

.1 소화 작업을 개시하시오.

.1.1 하나/둘/ …의 소화반/ …반을 현장에 보내시오.

.2 다음의 통로를 따라 가시오:

.2.1 기관실/제 …번 화물창/ 탱크/ 상부구조물/ 거주구역/ …장소/ …장소/ 연돌/ …의 맨홀을 통하여 가시오.

.2.2 바깥쪽/ 안쪽으로부터 …로 가시오.
좌현측/ 우현측으로부터 …로 가시오.
…로부터 …로 가시오.

.3 다음과 같은 (추가의)안전 조치를 취한 후 보고하시오.

.3.1 한 개의 팀에 2명/ …명의 구성원을 넣으시오.

.3.1.1 소화반 /…반 내의 구성원의 수는: …임.

.3.2 상호간에 / 외부와 구명줄을 연결하시오.

.3.2.1 …반의 구성원 사이에는 서로 구명줄을 연결하였음.

.3.2.2 …반은 외부와 구명줄을 연결하였음.

.3.3 Have rescue team on stand by.
.3.4 Maintain visual contact/radio contact on walkie-talkie.
.4 Fire fighting team must have following outfit:
.4.1 Fire fighting team must have protective clothing/smoke helmets /breathing apparatus/… .
.5 Manning of fire fighting team/… team(s) as follows:
.5.1 Chief Officer/Chief Engineer/… in command of fire fighting team/… team (no. …).
.5.2 Following officer(s)/crew member(s) in fire fighting team/… team: … .
.6 Restrict action (in …/ on …) to … minutes.
.6.1 Agree on retreat signal and report.
.6.1.1 Retreat signal for fire fighting team/… team is … .
.7 Use water/foam/powder/CO_2/sand/… in … .
.8 Run out fire hoses and report.
.8.1 Fire hoses run out.
.9 Water on!
.9.1 Water is on.
.10 Cool down … with water and report.
.10.1 … cooled down.

.4 Cancellation of alarm

.1 Is the fire extinguished?
.1.1 Yes, fire (in …) extinguished.
.1.2 No, fire (in …) not extinguished (yet).
.1.3 Fire restricted to … space/area.
.2 Post a fire watch and report.
.2.1 Fire watch posted (in …space/area).
.3 Fire extinguishing systems/means remain on stand-by.
.4 Fire fighting team/… team remain on stand-by.
.5 Rope-off the fire area and report.

.3.3 구조반을 대기하시오.
.3.4 시각통신을 유지하시오/워키토키로 무선통신을 유지하시오.
.4 소화반은 다음의 장구를 갖춰야함:
.4.1 소화반은 보호복/방연 헬멧/자장식 호흡구/ …을 갖춰야함.
.5 소화반/…반의 인적 구성은 다음과 같음:
.5.1 일등항해사/기관장/…가 (제…번)소화반/…반을 지휘함.
.5.2 소화반/…반 내에는 다음의 사관/선원이 속해 있음: … .
.6 (…내에서/…위에서) …분간으로 행동을 제한하시오.
.6.1 철수 신호에 합의하고 보고하시오.
.6.1.1 소화반/…반에 대한 철수 신호는 …임.
.7 …에 소화수/포말/분말/CO_2/모래/…을 사용하시오.
.8 소화 호스를 신출하고 보고하시오.
.8.1 소화 호스가 신출되었음.
.9 소화수를 보내시오!
.9.1 소화수를 보냈음.
.10 …을 물로 식히고 보고하시오.
.10.1 …을 물로 식혔음.

.4 경보의 해제

.1 화재는 소화되었는가?
.1.1 예, (…의)화재는 소화되었음.
.1.2 아니오, (…의)화재는 (아직) 소화되지 않았음.
.1.3 화재는 …장소/구역에 제한됨.
.2 화재 당직을 배치하고 보고하시오.
.2.1 (…장소/구역에)화재 당직을 배치하였음.
.3 소화 설비/소화 도구는 대기 상태로 있음.
.4 소화반/…반은 대기 상태로 있음.
.5 화재 구역을 로프로 격리하고 보고하시오.

.5.1 Fire area roped-off.

.6 Check the fire area every … minutes/hour(s) for reignition and report.

.6.1 Fire area checked, no re-ignition.

.6.2 Fire area checked, re-ignition in … space/area.

.6.2.1 Re-ignition extinguished.

.7 The fire alarm is cancelled (with following restrictions: ….)

B2/4 Damage Control

Also see B2/1 "General Activities".

B2/4.1 Checking equipment status and drills

.1 Check the openings in all spaces/in … and report.

.1.1 All openings in … are closed.

.1.2 Openings in … are not closed (yet).

.1.3 Openings in … are not accessible.

.2 Check the watertight door control and report.

.2.1 Watertight door control

~ is operational.

~ (in …) is not operational (yet).

~ (in …) will be operational in … minutes.

.2.2 Watertight door(s) (in …) is/are not accessible.

.3 Check the pumps/emergency generator and report

.3.1 (Bilge) pump(s) in …/emergency generator

~ is/are operational.

~ is/are not operational (yet).

~ will be operational in … minutes.

.4 Check the power supply and report

.4.1 Power (in/at …)

~ is available.

~ is not available (yet).

~ will be available in … minutes.

.5.1 화재 구역을 로프로 격리하였음.
.6 재발화에 대비하여 …분/시간마다 화재 구역을 점검하고 보고하시오.
.6.1 화재 구역을 점검하였음, 재발화는 없음.
.6.2 화재 구역을 점검하였음…장소/구역에 재발화 하였음.
.6.2.1 재발화는 소화되었음.
.7.1 (다음의 제한 하에서:…)화재 경보는 취소되었음.

B2/4 손상 제어

(B2/1 "일반 사항"도 참조)

B2/4.1 설비 상태의 점검과 훈련

.1 모든 장소/…내의 개구를 점검하고 보고하시오.
.1.1 …에 있는 모든 개구는 닫혔음.
.1.2 …에 있는 개구를 (아직) 닫히지 않았음.
.1.3 …에 있는 개구에는 접근할 수 없음.
.2 수밀문 제어기를 점검하고 보고하시오.
.2.1 수밀문 제어기는 작동 가능함.
(…에 있는)수밀문 제어기에는 (아직)작동 되지 않음.
(…에 있는)수밀문 제어기에는 …분 후 작동 가능함.
.2.2 (…에 있는)수밀문 제어기는 접근할 수 없음.
.3 펌프/비상 발전기를 점검하고 보고하시오.
.3.1 …의 (빌지)펌프/비상 발전기는 작동 가능함.
…의 (빌지)펌프/비상 발전기는 (아직)작동 되지 않음.
…의 (빌지)펌프/비상 발전기는 …분 후 작동 가능함.
.4 전원을 점검하고 보고하시오.
.4.1 (…내의/…의)전원은 사용 가능함.
(…내의/…의)전원은 (아직)사용 가능하지 않음.
(…내의/…의)전원은 …분 후 사용 가능함.

.5 Check the damage control equipment and report.
.5.1 All damage control equipment is complete and available.
.5.2 Damage control equipment is not complete.
.5.2.1 Complete the damage control equipment.

B2/4.2 Damage control activities

.1 Reporting flooding

.1 We have collided (with …) .
.2 We have flooding in … .
.3 Is flooding under control?
.3.1 Yes, flooding under control.
.3.2 No, flooding (in …) not under control (yet).
.4 Is danger imminent?
.4.1 No, danger not imminent.
.4.2 Yes, danger of (total) blackout (in …).
.4.3 Yes, danger of heavy listing/capsizing/sinking/… .

.2 Reporting readiness for action

.1 Muster damage control team and report.
.1.1 Damage control team stand complete and mustered.
.2 Is damage control material available?
.2.1 Yes, damage control material available.
.2.2 No, damage control material not available (yet).
.2.3 Damage control material will be available in … minutes.
.3 Stand by engine room/… station and report.
.3.1 Engine room/… station standing by.
.3.2 Engine room/… station flooded.
.3.3 Engine room/… station will be standing by in … minutes.

.5 손상 제어 장비를 점검하고 보고하시오.
.5.1 모든 손상 제어 장비는 완전하고 사용 가능함.
.5.2 손상 제어 장비는 완전하지 않음.
.5.2.1 손상 제어 장비를 완전하게 하시오.

B2/4.2 손상 제어 활동

.1 침수의 보고

.1 본선은 (…와)충돌하였음.
.2 본선은 …에 침수가 있음.
.3 침수는 진압되고 있는가?
.3.1 예, 침수는 진압되고 있음.
.3.2 아니오, (…안의)침수는 (아직)진압되지 않고 있음.
.4 위험이 급박한가?
.4.1 아니오, 위험은 급박하지 않음.
.4.2 예, (…에)(전체적인)정전의 위험이 있음.
.4.3 예, 큰 횡경사/전복/침몰/…의 위험이 있음.

.2 작업 준비 상황의 보고

.1 손상 제어반을 소집하고 보고하시오.
.1.1 손상 제어반은 완전한 상태이고 소집되었음.
.2 손상 제어용 재료는 사용 가능한가?
.2.1 예, 손상 제어용 재료는 사용 가능함.
.2.2 아니오, 손상 제어용 재료는 (아직) 사용 가능하지 않음.
.2.3 손상 제어용 재료는 …분 후 사용 가능할 것임.
.3 기관실/…부서는 대기하고 보고하시오.
.3.1 기관실/…부서는 대기중임.
.3.2 기관실/…부서는 침수되었음.
.3.3 기관실/…부서는 …분 후 대기하겠음.

.3 Orders for damage control

.1 Close all openings/outlets/valves (in …) and report.
.1.1 All openings/outlets/valves (in …) closed.
.1.2 Openings/outlets/valves in … not accessible/not operational.
.2 Switch on/off power (at/on/in …) and report.
.2.1 Power (at/on/in …) switched on/off.
.2.2 Power supply (at/on/in …) not operational.
.3 Close watertight door(s) (in …) (by hand) and report.
.3.1 Watertight door(s) (in …) closed.
.3.2 Watertight door(s) (in …) not accessible/not operational.
.4 Switch on (bilge) pump(s) (in …)and report.
.4.1 (Bilge) pump(s) (in …) switched on.
.4.2 (Bilge) pump(s) (in …) not operational.
.5 Switch over (bilge) pump(s) from … to … .
.5.1 (Bilge) pump(s) switched over.
.5.2 Switching over (bilge) pump(s) not possible.
.6 Start damage control.
.6.1 Take one/two/… damage control team(s) to scene.
.7 Go following route: … .
.7.1 Go through engine room/no. … hold(s)/tank(s)/superstructure /manhole/… space/… deck/… .
.8 Go from
~ outside/inside to … .
~ port side/starboard side to … .
~ … to … .
.9 Take following (additional) safety measures and report.
.9.1 Have two/… members in one damage control team.
.9.2 Have lifeline between each other/to outside.
.9.3 Have rescue team on stand by and report.
.9.3.1 Rescue team standing by.

.3 손상 제어 관련 명령

.1 (…의) 모든 개구/배출구/밸브를 폐쇄하고 보고하시오.
.1.1 (…의)모든 개구/배출구/밸브를 폐쇄하였음.
.1.2 …의 개구/배출구/밸브는 접근 불능/작동 되지 않음.
.2 (…의/…상의/…내의) 전원 켜고/끄고 보고하시오.
.2.1 (…의/…상의/…내의) 전원을 켰음/껐음.
.2.2 (…의/…상의/…내의) 전원 공급 장치는 작동 되지 않음.
.3 (…내의)수밀문을 (손으로)닫고 보고하시오.
.3.1 (…내의)수밀문을 닫았음.
.3.2 (…내의)수밀문은 접근 불가능함/작동 되지 않음.
.4 (…내의)(빌지)펌프에 전원을 켜고 보고하시오.
.4.1 (…내의)(빌지)펌프에 전원을 켰음.
.4.2 (…내의)(빌지)펌프는 작동 되지 않음.
.5 (빌지)펌프를 …에서…으로 전환하시오.
.5.1 (빌지)펌프를 전환하였음.
.5.2 (빌지)펌프를 전환할 수 없음.
.6 손상 제어를 개시하시오.
.6.1 손상 제어반 1개/2개/…를 현장에 보내시오.
.7 다음의 통로를 따라가시오: … .
.7.1 기관실/제…번 화물창/ 탱크/ 상부 구조물/ 맨홀/ …장소/… 갑판/ …을 통하여 가시오.
.8 바깥쪽/안쪽으로부터…으로 가시오.
좌현쪽/우현쪽으로부터 …으로 가시오.
…으로부터 …으로 가시오.
.9 다음과 같은 (추가의)안전 조치를 취하고 보고하시오.
.9.1 하나의 손상 제어반에 2명/…명의 인원을 배치하시오.
.9.2 구명줄을 서로 간에/외부와 연결하시오.
.9.3 구조반을 대기하고 보고하시오.
.9.3.1 구조반은 대기 중임.

.9.4 Maintain visual contact/radio contact on walkie-talkie.
.10 Damage control team must have following outfit(s).
.10.1 Damage control team must have
~ protective clothing.
~ safety helmets.
~ lifejackets.
~ diving equipment/… .
.11 Manning of damage control team as follows: … .
.11.1 Chief Officer/Chief Engineer/… in command of damage control team (no. …) … .
.11.2 Following officer(s)/crew member(s) in damage control team (no. …): … .
.12 Restrict action (in …) to … minutes.
.12.1 Agree on retreat signal and report.
.12.1.1 Retreat signal … .
.13 Stop flooding from inside/outside (… space/area) and report.
.13.1 Flooding stopped.
.13.2 Stopping flooding from inside/outside not possible.

.4 Cancellation of alarm

.1 Has flooding stopped?
.1.1 Yes, flooding (in …) has stopped.
.1.2 No, flooding (in …) has not (completely) stopped (yet).
.2 Is flooding under control?
.2.1 Yes, flooding (in …) under control.
.2.2 Flooding (in …) below/above capacity of (bilge) pump(s).
.2.3 Flooding restricted to … space/area.
.3 Post damage control watch and report.
.3.1 Damage control watch posted (in …).
.4 How much water is in the vessel?
.4.1 Quantity of water (in …) about … tonnes.

.9.4 시각통신을 유지하시오./워키토키로 무선통신을 유지하시오.
.10 손상 제어반은 다음의 장구를 구비하여야 함.
.10.1 손상 제어반은 보호복을 구비하여야 함.
손상 제어반은 안전 헬멧을 구비하여야 함.
손상 제어반은 구명동의를 구비하여야 함.
손상 제어반은 잠수 기구/…를 구비하여야 함.
.11 손상 제어반의 인원 배치는 다음과 같음:… .
.11.1 일등항해사/기관장/ …가(제 …번)손상 제어반을 지휘함.
.11.2 다음의 사관/승무원이(제 …번)손상 제어반에 속함: … .
.12 (…에서는) …분 간으로 행동을 제한하시오.
.12.1 철수 신호에 대하여 합의하고 보고하시오.
.12.1.1 철수 신호는 …임.
.13 (…장소/구역의) 안쪽/바깥쪽으로부터 누수를 차단하고 보고하시오.
.13.1 누수가 차단되었음.
.13.2 안쪽/바깥쪽으로부터의 누수 차단이 불가능함.

.4 경보의 해제

.1 침수는 중지되었는가?
.1.1 예, (…내의) 침수는 중지되었음.
.1.2 아니오, (…내의) 침수는 (아직) (완전히) 중지되지 않았음.
.2 침수는 진압되고 있는가?
.2.1 예, (…내의)침수는 진압되고 있음.
.2.2 (…내의)침수는 (빌지)펌프의 용량 이내/이상임.
.2.3 침수는 …장소/구역 내로 한정되었음.

.3 손상 제어 당직을 배치하고 보고하시오.
.3.1 (…내에) 손상 제어 당직이 배치되었음.
.4 선박 내에 물이 얼마나 있는가.
.4.1 (…내의)물의 양은 대략…톤임.

.4.2 Quantity of water (in …) not dangerous.
.5 (Bilge) pump(s) remain on stand by.
.6 Engine room remains on stand by.
.7 Additional emergency generator remains on stand by.
.8 Damage control team remains on stand by.
.9 Rope-off flooded area.
.10 Check leak every … minutes/hour(s) and report.
.10.1 Leak checked - no flooding.
.10.2 Leak checked - minor/major flooding (in …).
.10.2.1 Flooding has stopped.
.11 The alarm is cancelled (with following restrictions: …).

B2/5 Grounding

Also see B2/1 "General Activities"

B2/5.1 Reporting grounding and ordering actions

.1 We are aground.
.2 Stop engine(s).
.3 Close watertight doors and report.
.3.1 Watertight doors closed.
.4 Is vessel (still) making way?
.4.1 Yes, vessel making way ahead/astern.
.4.2 No, vessel not making way.
.5 Give "vessel aground" signals.[102)]
.6 Inform engine room.
.7 What part is aground?
.7.1 Vessel aground forward/amidships/aft/full length.
.8 Stand by forward station and aft station and report.
.8.1 Forward station/aft station standing by.
.9 Stand by port anchor/starboard anchor.

.4.2 (…내의)물의 양은 위험하지 않음.
.5 (빌지)펌프는 대기 상태임.
.6 기관실은 대기 상태로 있음.
.7 추가의 비상 발전기는 대기 상태로 있음.
.8 손상 제어반은 대기 상태로 있음.
.9 침수 구역을 로프로 격리하시오.
.10 누수를 …분/시간마다 점검하고 보고하시오.
.10.1 누수를 점검하였음 - 침수는 없음.
.10.2 누수를 점검하였음 - (…에)소량의/다량의 침수가 있음.
.10.2.1 침수는 중지되었음.
.11 (다음의 제한 사항과 함께:…) 경보는 취소되었음.

B2/5 좌초

(B2/1"일반 사항"도 참조)

B2/5.1 좌초의 보고 및 조치의 명령

.1 본선은 좌초하였음.
.2 기관을 정지하시오.
.3 수밀문을 폐쇄하고 보고하시오.
.3.1 수밀문은 폐쇄되었음.
.4 본선은 (아직도) 타력이 있는가?
.4.1 예, 본선은 전진/후진 타력이 있음.
.4.2 아니오, 본선은 타력이 없음.
.5 "선박 좌초" 신호를 하시오.
.6 기관실에 통지하시오.
.7 어느 부분이 좌초되었는가?
.7.1 본선의 선수/중앙/선미/전체가 좌초되었음.
.8 선수 부서와 선미 부서는 대기하고 보고하시오
.8.1 선수 부서/선미 부서는 대기하고 있음
.9 좌현/우현 묘를 대기하시오

.10 What is position?
.10.1 Position … .

B2/5.2 Reporting damage

.1 Report damage.
.1.1 No damage.
.1.2 Crack(s) in plating/no. … double bottom/no. … hold(s)/ tank(s) /main/auxiliary engine(s) foundation/… .
.1.3 Deformation(s)/indentation(s) to plating/to … .
.1.4 Deformation(s)/indentation(s) to … .
.2 Check flooding and report.
.2.1 No flooding.
.2.2 Flooding in … .
.3 Is danger imminent?
.3.1 No, danger not imminent.
.3.2 Yes, danger of
~ heavy listing (to port/starboard)
~ decreasing stability.
~ damage by sea.
~ breaking apart.
~ environmental pollution.
~ … .
.4 What is nature of sea bottom?
.4.1 Sea bottom rocky.
.4.2 Sea bottom soft.
.5 What is state of tide?
.5.1 No tide.
.5.2 Tide … metres/rising/falling/turning at … UTC/within … hours.
.6 What is wind force and direction?
.6.1 Wind force Beaufort… from …*(cardinal points).*
.6.1.1 Wind expected to decrease/increase (within the next … hours).

.10 위치는 어디인가?
.10.1 위치는 …임

B2/5.2 손상의 보고

.1 손상을 보고하시오.
.1.1 손상 없음
.1.2 철판/ 제 …번 이중저/ 제 …번 화물창/ 탱크/ 주기관/ 보조기관의 기반/ …에 금이 갔음
.1.3 철판/…에 변형/움푹 들어간 곳이 있음
.1.4 …에 변형/움푹 들어간 곳이 있음
.2 침수를 점검하고 보고하시오
.2.1 침수가 없음
.2.2 …에 침수가 있음.
.3 위험이 급박한가?
.3.1 아니오, 위험이 급박하지 않음
.3.2 예, (좌현/우현으로) 크게 횡경사할 위험이 있음.
예, 복원성이 감소될 위험이 있음
예, 파도에 의하여 손상될 위험이 있음
예, 절단될 위험이 있음
예, 환경 오염의 위험이 있음
예, …의 위험이 있음
.4 해저의 저질은 무엇인가?
.4.1 해저의 저질은 바위임.
.4.2 해저의 저질은 부드러움.
.5 조석의 상태는 어떤가?
.5.1 조석은 없음.
.5.2 조석은 세계시…시에/ …시간 이내에 …미터임/ 상승중임/하강중임/ 변하고 있음.
.6 풍력과 풍향은 어떠한가?
.6.1 풍력계급은 … 풍향은 …(방점)임.
.6.1.1 (앞으로 …시간 내에) 바람은 감소/증가할 것으로 예상됨.

.6.1.2 Wind expected to back/veer (within the next ··· hours).
.6.1.3 No change expected (within the next ··· hours).
.7 What is sea state?
.7.1 Sea smooth/moderate/rough/high/swell slight/moderate/heavy ··· metres from ···*(cardinal points).*
.7.2 Sea smooth/moderate/rough/high/swell slight/moderate/heavy expected to decrease/increase (within the next ··· hours).
.7.3 No change expected (within the next ··· hours).
.8 What is draft?
.8.1 Draft ··· metres (port side/starboard side) forward/aft/amidships.
.9 What is depth of water?
.9.1 Greatest depth ··· metres (port side/starboard side) forward/aft/amidships.

B2/5.3 Orders for refloating

.1 Are (bilge) pumps operational?
.1.1 Yes, (bilge) pumps operational.
.1.2 No, (bilge) pumps not operational (yet).
.1.3 (Bilge) pumps will be operational in ··· minutes.
.2 Is damage control material available?
.2.1 Yes, damage control material available.
.2.2 No, damage control material not available (yet).
.2.3 Damage control material will be available in ··· minutes.
.3 Stand by engine room and report.
.3.1 Engine room standing by.
.4 Stand by all anchors for letting go.
.5 Report distribution of cargo.
.5.1 No. ··· hold(s)/tank(s) ··· tonnes (of ··· cargo).
.5.2 Deck cargo foreward/aft/amidships ··· tonnes (of ···).
.5.3 Forepeak/afterpeak ··· tonnes.
.5.4 No. ··· double bottom tank(s) ··· tonnes (of ballast/···).

.6.1.2 (앞으로 …시간 이내에) 풍향은 반전/순전할 것으로 예상됨.
.6.1.3 (앞으로 …시간 이내에) 변화가 없을 것으로 예상됨.
.7 해면 상태는 어떤가?
.7.1 파도는 잔잔함/보통/거침/높음 너울은 약간임/보통임/심함, …미터이고 …(방점)에서 오고 있음.
.7.2 (앞으로 …시간 이내에)잔잔한/ 보통인/ 거친/ 높은 파도, 약간의/보통의/심한 너울은 감소/증가할 것으로 예상됨.
.7.3 (앞으로 …시간 이내에) 변화가 없을 것으로 예상됨.
.8 흘수는 얼마인가?
.8.1 (좌현측/우현측)선수/선미/중앙부 흘수는 …미터임.
.9 수심은 얼마인가?
.9.1 (좌현측/우현측)선수/선미/중앙부의 최대 수심은 …미터임.

B2/5.3 재부양 작업 관련 명령

.1 (빌지)펌프는 작동 가능한가?
.1.1 예, (빌지)펌프는 작동 가능함.
.1.2 아니오, (빌지)펌프는 (아직) 작동 되지 않음.
.1.3 (빌지)펌프는 …분 후 작동 가능할 것임.
.2 손상 제어용 재료는 사용 가능한가?
.2.1 예, 손상 제어용 재료는 사용 가능함.
.2.2 아니오, 손상 제어용 재료는 (아직) 사용 가능하지 않음.
.2.3 손상 제어용 재료는 …분 후 사용 가능할 것임.
.3 기관실을 대기시키고 보고하시오.
.3.1 기관실은 대기중임.
.4 모든 닻을 투묘 준비하시오.
.5 화물 배치 상태를 보고하시오.
.5.1 제…번 화물창/탱크에 (…화물)…톤이 있음.
.5.2 갑판적 화물은 선수부/선미부/중앙부에(…화물) …톤임
.5.3 선수/선미 탱크에 …톤이 있음.
.5.4 제…번 이중저 탱크에 (밸러스트/…)…톤이 있음.

.6 Transfer cargo from no. … hold(s)/tank(s) to no. … hold(s)/ tank(s) and report.
.6.1 Cargo from no. … hold(s)/tank(s) transferred to no. … hold(s) /tank(s).
.7 Transfer deck cargo from … to … and report.
.7.1 Deck cargo from … transferred to … .
.8 Pump out forepeak/afterpeak and report.
.8.1 Forepeak/afterpeak pumped out.
.9 Transfer ballast/… from no. … double bottom tank(s) to no. … double bottom tank(s) and report.
.9.1 Ballast/… from no. … double bottom tanks transferred to no. … double bottom tank(s).
.10 Fill forepeak/afterpeak.
.11 Jettison cargo from … and report.
.11.1 Cargo from … jettisoned .
.12 Engine(s) full/… astern/ahead.
.13 Has vessel refloated?
.13.1 Yes, vessel refloated.
.13.2 No, vessel not refloated (yet).

B2/5.4 Checking seaworthiness

.1 Request a (diving) survey.
.2 Report the result of the (diving) survey.
.2.1 No damage.
.2.2 Following damage to the plating: … .
.2.2.1 Crack(s) in area of … .
.2.2.2 Deformation(s)/indentation(s) in area of … .
.2.3 Following damage to the engine(s)/pipe(s): … .
.2.3.1 Crack(s) in the main engine(s)/auxiliary engine(s) foundation.
.2.3.2 Deformations/fracture(s) to the pipe(s) in/out … .
.2.3.3 Fractures/bending of the bolt(s) of … .

.6 제…번 화물창/탱크로부터 제…번 화물창/탱크로 화물을 이송하고 보고하시오.

.6.1 제…번 화물창/탱크로부터 제…번 화물창/탱크로 화물이 이송되었음.

.7 갑판적 화물을 …에서 …으로 이송하고 보고하시오.

.7.1 갑판적 화물이 …에서 …으로 이송되었음.

.8 선수/선미 탱크를 배출하고 보고하시오.

.8.1 선수/선미 탱크를 배출하였음.

.9 제…번 이중저 탱크에서 제…번 이중저 탱크로 밸러스트/ … 을 이송하고 보고하시오.

.9.1 밸러스트/…가 제…번 이중저 탱크에서 제…번 이중저 탱크로 이송 되었음.

.10 선수/선미 탱크를 채우시오.

.11 …로부터 화물을 투하하고 보고하시오.

.11.1 …로부터 화물이 투하되었음.

.12 기관을 전속/… 후진/전진하시오.

.13 선박은 재부양하였는가?

.13.1 예, 선박은 재부양하였음.

.13.2 아니오, 선박은 (아직) 재부양하지 않았음.

B2/5.4 감항성 점검

.1 (잠수)검사를 요청하시오.

.2 (잠수)검사의 결과를 보고하시오.

.2.1 손상은 없음.

.2.2 외판에 다음의 손상이 있음:…

.2.2.1 …구역에 금이 갔음.

.2.2.2 …구역에 변형/움푹 들어간 곳이 있음.

.2.3 기관/파이프에 다음의 손상이 있음:…

.2.3.1 주기관/보조기관의 기반에 금이 갔음.

.2.3.2 …내의/로부터의 파이프에 변형/파열이 있음.

.2.3.3 …의 볼트에 파손/굴곡이 있음.

.2.4 Following damage to the underwater hull: (also see: .1.2.2)
.2.4.1 Deformation(s)/indentation(s) to the sea water inlet(s)/outlet(s).
.2.4.2 Deformation(s)/indentation(s) to the stem/bulb.
.2.4.3 Deformation(s) to the propeller(s).
.2.4.4 (Port/starboard) propeller(s) missing.
.2.4.5 Deformation to the rudder/to … .
.2.5 Dry-docking is recommended/necessary.
.3 Is the vessel seaworthy?
.3.1 Yes, the vessel is seaworthy.
.3.2 No, the vessel is not seaworthy (yet).
.3.2.1 The vessel must be repaired and re-inspected.
.3.3 Request … tug(s).

B2/6 Search and rescue on-board activities

For details also see IAMSAR-Manual, London/Montreal, 1998.[103)]

B2/6.1 Checking equipment status

.1 Check the lifebuoys and report.
.1.1 All lifebuoys are complete.
.1.2 Lifebuoy(s) at … is/are damaged/missing.
.1.2.1 Replace the damaged/missing lifebuoy(s).
.2 When was the last man overboard drill?
.2.1 Last man overboard drill was on …*(date).*
.3 Prepare a plan for man overboard drill.
.3.1 Prepare a plan for
~ an announced/not announced drill.
~ a daytime/nighttime drill.
~ a muster (at all stations).
~ a recovering manoeuvre (with dummy/buoy).
.4 Have a drill/manoeuvre/muster on …*(date).*

.2.4 수선 하부의 선체에 다음의 손상이 있음:(.1.2.2참조)
.2.4.1 해수의 흡입구/배출구에 변형/움푹 들어간 곳이 있음.
.2.4.2 선수재/구형 선수에 변형/움푹 들어간 곳이 있음.
.2.4.3 프로펠러에 변형이 있음.
.2.4.4 (좌현/우현 측)프로펠러가 없어졌음.
.2.4.5 타/…에 변형이 있음.
.2.5 입거가 권고됨/필요함.
.3 선박은 감항성이 있는가?
.3.1 예, 선박은 감항성이 있음.
.3.2 아니오, 선박은 (아직)감항성이 없음.
.3.2.1 선박은 수리되고 다시 검사되어야함.
.3.3 예인선 …척을 요청함.

B2/6 수색과 구조 관련 선내의 활동

(세부사항에 관하여는:IAMSAR지침 참조)

B2/6.1 설비 상태의 점검

.1 구명부환을 점검하고 보고하시오.
.1.1 모든 구명부환은 완전함.
.1.2 …의 구명부환은 손상되었음/없어졌음.
.1.2.1 손상된/없어진 구명부환을 새로 설치하시오.
.2 지난번의 익수자 훈련은 언제 하였는가?
.2.1 지난번의 익수자 훈련은 …(날짜)에 하였음.
.3 익수자 훈련의 계획을 마련하시오.
.3.1 공고된/공고 안된 훈련의 계획을 마련하시오.
주간/야간 훈련의 계획을 마련하시오.
(모든 부서의)소집 계획을 마련하시오.
(표적/부표를 사용하여) 원침로로 찾아가는 조종 훈련의 계획을 마련하시오.
.4 …(날짜)에 훈련/조종/소집을 실시하시오.

B2/6.2 Person-overboard activities

.1 Man overboard (on port side/starboard side/astern)!
.2 Drop lifebuoy(s).
.2.1 Sound "man overboard" alarm.
.3 Hoist flag signal "Oscar".
.4 Hard-a-port/hard-a-starboard the wheel.
.5 Is person in water/lifebuoy located?
.5.1 Yes, person in water/lifebuoy located.
.5.2 Report direction and distance of person in water/lifebuoy.
.5.2.1 Direction at … points/degrees port side/starboard side, distance … metres.
.5.2.2 Maintain visual contact to person in water/lifebuoy.
.5.3 No, person in water/lifebuoy not located (yet).
.5.3.1 Look out for person in water/lifebuoy and report.
.5.4 Passenger/crew member missing (for … hours/since … UTC) - search in vessel negative.
.5.4.1 Stop engine(s).
.5.4.2 Transmit alarm signal - PAN-PAN/distress alert-MAYDAY to coast radio station/Maritime Rescue Co-ordination Centre/ vessels in vicinity and report.
.5.4.3 Alarm signal - PAN-PAN/distress alert - MAYDAY transmitted /acknowledged by …/not acknowledged (yet).
.6 Return manoeuvre! Port/Starboard, steer … degrees.
.7 Report position.
.7.1 Position … .
.8 Report traffic situation.
.8.1 No vessel in vicinity.
.8.2 Following vessel(s) in vicinity: … .
.9 Report weather situation.
.9.1 Sea smooth/moderate/rough/high - swell slight/moderate /heavy from … *(cardinal points)*.

B2/6.2 익수자 관련 활동

.1 (좌현측/우현측/선미에서) 사람이 물에 빠졌음!
.2 구명부환을 던지시오.
.2.1 "사람이 물에 빠졌음"의 음향 경보를 발하시오.
.3 "Oscar"신호기를 게양하시오.
.4 타륜을 좌현측/우현측으로 완전히 돌리시오.
.5 물속의 사람/구명부환의 위치를 확인했는가?
.5.1 예, 물속의 사람/구명부환의 위치를 확인했음.
.5.2 물속의 사람/구명부환의 방향과 거리를 보고하시오.
.5.2.1 방향은 좌현측/우현측…점/…도, 거리는 …미터임.
.5.2.2 물속의 사람/구명부환과 시각적 접촉을 유지하시오.
.5.3 아니오, 물속의 사람/구명부환은 (아직)위치를 확인하지 못했음.
.5.3.1 물속의 사람/구명부환에 대하여 견시하고 보고하시오.
.5.4 (…시간 동안/세계시…시 이래)여객/승무원이 없어졌음.
- 선내의 수색 결과는 부정적임.
.5.4.1 기관을 정지하시오.
.5.4.2 경보신호-팡 팡/조난경보-메이데이를 해안 무선국/해상 구조 조정본부/근처의 선박에게 발신하고 보고하시오.
.5.4.3 경보신호-팡 팡/조난경보-메이데이를 발신하였음/…으로 부터 수신확인이 되었음/(아직)수신 확인되지 않았음.
.6 회귀 조종! 좌현 전타/우현 전타, …도로 조종하시오.
.7 위치를 보고하시오.
.7.1 위치는 …임.
.8 통항 상황을 보고하시오.
.8.1 근처에 선박이 없음.
.8.2 다음의 선박이 근처에 있음: … .
.9 기상 상태를 보고하시오.
.9.1 파도는 잔잔함/보통임/거침/높음 - 너울은 약간임/보통임/심함이고 …(방점)에서 옴.

.9.2 Winds force Beaufort … from … *(cardinal points).*
.9.3 Visibility good/moderate/poor.
.9.4 Current … knots to … *(cardinal points).*
.10 Have man overboard stations/lookouts at … manned and report.
.10.1 Man overboard stations/lookouts at … manned.
.11 Stand by for recovering from shipboard and report.
.11.1 Standing by for recovering from shipboard.
.12 Stand by rescue boat/motor lifeboat no. … for letting go and report.[104)]
.12.1 Rescue boat/motor lifeboat no. … standing by for letting go.
.13 Let go rescue boat/motor lifeboat.
.14 Use VHF Channel …/frequency … for communication.
.14.1 Use light signals/flag signals/whistle for communication.
.15 What is retreat signal for rescue boat/motor lifeboat?
.15.1 Retreat signal … .
.16 Stand by one/two crew member(s) for rescue in water and report.
.16.1 One/two crew member(s) standing by for rescue in water.
.17 Person overboard rescued/recovered.
.18 Stand by boat/rescue litter/rescue net/rescue basket/ rescue sling and report.
.18.1 Boat/rescue litter/rescue net/rescue basket/rescue sling standing by.
.19 Hoist person and report.
.20 Report condition of survivor.
.20.1 Survivor
~ is in good/bad condition.
~ has hypothermia.
~ is injured.
~ is suffering from shock.
.20.2 Person is dead.

.9.2 바람 상태는 풍력계급…이고 …(방점)으로부터 불어옴.
.9.3 시정은 양호함/보통임/나쁨.
.9.4 해류는 …노트 …(방점) 방향으로 흐름.
.10 익수자 부서에/…의 견시대에 인원을 배치하고 보고하시오.
.10.1 익수자 부서에/…의 견시대에 인원을 배치하였음.
.11 현측에서 끌어올릴 준비를 하고 보고하시오.
.11.1 현측에서 끌어올릴 준비를 하였음.

.12 구조정/제…번 동력 구명정을 보낼 준비를 하고 보고하시오.
.12.1 구조정/제…번 동력 구명정을 보낼 준비를 하였음.
.13 구조정/ 동력 구명정을 보내시오.
.14 통신을 위하여 VHF 채널…/주파수…을 사용하시오.
.14.1 통신을 위하여 등화/신호기/기적을 사용하시오.
.15 구조정/동력 구명정에 대한 철수신호는 무엇인가?
.15.1 철수신호는 …임.
.16 물속에서의 구조를 위하여 한 명의/두 명의 승무원을 대기시 키고 보고하시오.
.16.1 물속에서의 구조를 위하여 한/두 명의 승무원을 대기시켰음.
.17 익수자는 구조되었음/끌어올려졌음.
.18 보우트/ 구조 들것/ 구조 네트/ 구조 바구니/ 구조 슬링을 대기하고 보고하시오.
.18.1 보우트/구조 들것/구조 네트/구조 바구니/구조 슬링을 대기하였음.
.19 사람을 끌어올리고 보고하시오.

.20 생존자의 상태를 보고하시오.
.20.1 생존자는 좋은/나쁜 상태임.
생존자는 저체온 증임.
생존자는 부상당했음.
생존자는 충격을 받았음.
.20.2 사람은 사망하였음.

B2/6.3 Rescue operation-reporting readiness for assistance

Also see AI/1.2 "Search and rescue communications"

.1 Received an alarm signal- PAN-PAN/distress alert-MAYDAY at … UTC on … *(VHF Channel/frequency).*
.2 Observed the following distress signal in … degrees.
.3 Report the distress position.
.3.1 Distress position … .
.4 Was the alarm signal- PAN-PAN/distress alert - MAYDAY acknowledged?
.4.1 The alarm signal- PAN-PAN/distress alert - MAYDAY was acknowledged by …/not acknowledged (yet).
.4.2 Acknowledge the PAN-PAN/distress alert - MAYDAY - RELAY.
.4.3 Transmit a MAYDAY - RELAY to … *(radio station).*
.5 Watch the radar.
.6 Have the lookouts manned and report.
.6.1 Lookouts are manned.
.7 Contact vessels in vicinity of the distress and report.
.7.1 We have contact to following vessel(s) in vicinity of the distress:… .
.7.2 We have no contact (yet).
.8 Request information from the vessel in distress and report.
.8.1 We have following information from the vessel in distress: … .
.8.2 We have no information (yet).
.9 Stand by lines/lifebuoys/nets/derricks/cranes/… and report.
.9.1 Lines/lifeboats/nets/derricks/cranes/… standing by.
.10 Stand by lifeboats/rescue boat and report.
.10.1 Lifeboats/rescue boat standing by.
.11 Stand by liferaft(s) as boarding station(s) and report.
.11.1 Liferaft(s) standing by as boarding station(s).

B2/6.3 구조 작업-원조 준비 상황의 보고
(AI/1.2 "수색과 구조 통신"도 참조)

.1 세계시 …시에 …(VHF 채널/주파수)에 의하여 경보신호- 팡-팡 / 조난 경보-메이데이를 수신하였음.
.2 …도 방향에서 다음의 조난신호를 관측했음.
.3 조난위치를 보고하시오.
.3.1 조난위치는 …임.
.4 경보신호-팡-팡/조난경보-메이데이는 수신 확인되었는가?
.4.1 경보신호-팡-팡/조난경보-메이데이는 …에 의하여 수신 확인되었음/(아직) 수신 확인되지 않았음.
.4.2 팡-팡/조난경보-메이데이-릴레이의 수신 확인을 하시오.
.4.3 메이데이-릴레이를 …(무선국)에 발신하시오.
.5 레이더를 주시하시오.
.6 견시대에 인원을 배치하고 보고하시오.
.6.1 견시대에 인원을 배치하였음.
.7 조난 위치 부근의 선박과 교신하고 보고하시오.
.7.1 이쪽은 조난 위치 부근의 다음의 선박들과 교신하였음: … .
.7.2 이쪽은(아직) 교신하지 못했음.
.8 조난 선박에게 정보를 요청하고 보고하시오.
.8.1 이쪽은 조난 선박으로부터 다음의 정보를 입수하였음 : … .
.8.2 이쪽은(아직)정보가 없음.
.9 밧줄/ 구명부환/ 네트/ 데리크/ 크레인/ …을 대기하고 보고 하시오.
.9.1 밧줄/ 구명부환/ 네트/ 데리크/ 크레인/ …을 대기하였음.
.10 구명정/ 구조정을 대기하고 보고하시오.
.10.1 구명정/ 구조정을 대기하였음.
.11 승선장으로서 구명벌을 대기하고 보고하시오.
.11.1 승선장으로서 구명벌을 대기하였음.

.12 Let go liferaft(s) as boarding station(s) with … crew members (each).
.13 Stand by … crew members for assisting survivors in water and report.
.13.1 … crew members standing by for assisting survivors in water.
.14 Switch on the deck lighting/outboard lighting/search lights.
.15 Stand by line throwing apparatus and report.
.15.1 Line throwing apparatus standing by.

B2/6.4 Conducting search

.1 We/MV … will act as On-scene Co-ordinator.
.1.2 Inform coast radio station(s)/MRCC/ vessels in vicinity.
.2 Stand by bridge team/lookouts for information/signals of On-scene Co-ordinator.
.2.1 Following information/signal received from On-scene Co-ordinator:
.3 We carry out search pattern …/radar search.
.3.1 We start search pattern …/radar search at … UTC.
.3.1.1 Inform the crew/lookouts/engine room.
.4 Bridge team/lookouts!
Keep sharp lookout for signals/sightings of the vessel in distress and report every … minutes.
.4.1 Light signals/smoke signals/sound signals/… signals in … degrees.
.4.2 Objects/vessel in distress/lifeboat(s)/life raft(s)/person(s) in water in … degrees.
.5 Stand by rescue team/boat crews/engine room and report.
.5.1 Rescue team/boat crews/engine room standing by.
.6 Transmit the following information/signals to the searching vessel(s):… .

.12 승선장으로서 승무원 …명(씩)을 태운 구명벌을 보내시오.

.13 물속의 생존자를 원조하기 위하여 …명의 승무원을 대기시키고 보고하시오.

.13.1 물속의 생존자를 원조하기 위하여 …명의 승무원을 대기 시켰음.

.14 갑판 조명등/선외 조명등/탐조등을 점등하시오.

.15 투삭 장비를 대기하고 보고하시오.

.15.1 투삭 장비를 대기하였음.

B2/6.4 수색 작업의 실시

.1 본선/기선…은 현장 지휘자로서 활동하겠음.

.1.2 해안 무선국/ 해상 구조 조정 본부/ 근처 선박에게 통보하시오.

.2 현장 지휘자의 통보/ 신호에 대비하여 선교팀/ 견시원을 대기하시오.

.2.1 현장 지휘자로부터 다음의 통보/신호가 수신되었음:

.3 우리는 수색 방식 …/레이더 수색을 수행함.

.3.1 우리는 세계시…시에 수색 방식…/레이더 수색을 시작함.

.3.1.1 승무원/견시원/기관실에 통보하시오.

.4 선교팀/견시원!
신호에 대비하여/ 조난선의 발견을 위하여 엄중한 견시를 유지하고 …분마다 보고하시오.

.4.1 등화 신호/ 연기 신호/ 음향 신호/…신호가 …도 방향에 있음.

.4.2 물체/ 조난선/ 구명정/ 구명벌/ 물 속의 사람이 …도 방향에 있음.

.5 구조반/보우트 승무원/기관실을 대기하고 보고하시오.

.5.1 구조반/보우트 승무원/기관실을 대기하였음.

.6 다음의 통보/신호를 수색선에게 발신하시오: … .

B2/6.5 Rescue activities

Also see B2/6.2 "Person-overboard activities"

.1 Rescue persons in following order:
- persons in water
- injured/helpless persons
- women and children
- passengers
- crewmembers.

.2 Ask the survivor(s) the following information:

.2.1 What was the total number of persons on board the vessel in distress?

.2.1.1 Total number of persons was: … .

.2.2 What was the number of casualties?

.2.2.1 Number of casualties was: … .

.2.3 What was the number of lifeboats/liferafts launched?

.2.3.1 Number of lifeboats/liferafts launched was: … .

.2.4 What was the number of persons in lifeboats/liferafts?

.2.4.1 Number of persons in lifeboats/liferafts was: … .

.2.5 What was the number of persons in water?

.2.5.1 Number of persons in water was: … .

.3 Inform … coast radio station about the name(s)/call sign(s) and destination of the vessel(s) with the survivors.

.3.1 Inform about the number of survivors on (each) vessel.

.3.2 Inform about the condition of the survivors.

.4 Inform … coast radio station about the condition of the vessel in distress.

.4.1 The vessel in distress
- ~ capsized/sunk/adrift (near position …)/drifting in … degrees.
- ~ grounded (in position …).
- ~ on fire.
- ~ not under command.

B2/6.5 구조 활동

(B2/6.2 "익수자 관련 활동"도 참조)

.1 다음의 순서로 사람을 구조하시오:
- -물속에 있는 자
- -부상자/거동이 불편한 자
- -여성과 어린이
- -여객
- -승무원

.2 생존자에게 다음의 정보를 요청하시오:

.2.1 조난선에 승선한 전체 인원수는 몇 명이었는가?

.2.1.1 전체 인원수는: …임.

.2.2 사상자의 수는 몇 명이었는가?

.2.2.1 사상자의 수는: …임.

.2.3 진수된 구명정/구명벌의 수는 몇이었는가?

.2.3.1 진수된 구명정/구명벌의 수는: …임.

.2.4 구명정/구명벌 내의 인원수는 몇 명이었는가?

.2.4.1 구명정/구명벌 내의 인원수는: …임.

.2.5 물에 빠진 사람의 수는 몇 명이었는가?

.2.5.1 물에 빠진 사람의 수는: …임.

.3 …해안 무선국에 생존자를 태운 선박의 선명/호출부호와 목적지를 통보하시오.

.3.1 (각)선박에 있는 생존자의 수를 알리시오.

.3.2 생존자의 상태에 관하여 알리시오.

.4 조난선의 상태에 대하여 …해안무선국에 알리시오.

.4.1 조난선은 전복됨/ 침몰됨/ (위치…부근에서)표류중임/ …도 방향으로 표류하고 있음.
조난선은 (위치…에서)좌초되었음.
조난선은 불이 났음.
조난선은 운전이 부자유함.

.5 Transmit the following safety message/navigational warning:
Vessel in distress(in position …) danger to navigation.

B2/6.6 Finishing with search and rescue operations

.1 Search and rescue finished at … UTC.
.1.1 Inform the crew/lookouts/engine room.
.2 We resume on-board routine at … UTC.
.3 Inform the coast radio station/searching vessels about the cancellation of search and rescue.
.4 We proceed with our voyage.

B3 Cargo and cargo handling

B3/1 Cargo handling

B3/1.1 Loading and unloading

.1 Loading capacities and quantities

.1 What is the deadweight of the vessel?
.1.1 The deadweight is … tonnes.
.2 What is the hold/bale/grain capacity of vessel?[105)]
.2.1 The hold/bale/grain capacity is … cubic metres.
.3 What is the container capacity of the vessel?
.3.1 The container capacity is … TEU.
.4 How many 20'/40' containers will the vessel load?
.4.1 The vessel will load … 20'/40' containers.
.5 How many cubic metres of cargo space are required?
.5.1 … cubic metres of cargo space are required.

.5 다음의 안전 통보/항행 경보를 발신하시오:
(위치…의)조난선은 항해상 위험함.

B2/6.6 수색 구조 작업의 종료.

.1 세계시 …시에 수색 구조 작업이 종료되었음.
.1.1 승무원/견시원/기관실에 통보하시오.
.2 우리는 세계시 …시에 선내 정상 과업을 재개함.
.3 수색 구조 작업의 취소에 대하여 해안무선국/수색선에 통보하시오.
.4 본선은 본선의 항해(계획)에 따라 항진함.

B3 화물 및 하역 작업

B3/1 하역 작업

B3/1.1 적하 및 양하

.1 적재 능력과 적재량

.1 본선의 재화중량은 얼마인가?
.1.1 재화중량은 …톤임.
.2 본선의 선창/ 베일/ 그레인 커패시티는 얼마인가?
.2.1 선창 용적/베일 커패시티/그레인 커패시티는 …입방미터임.
.3 본선의 컨테이너 커패시티는 얼마인가?
.3.1 컨테이너 커패시티는 …TEU임.
.4 본선에는 몇 개의 20'/40' 컨테이너를 적재할 것인가?
.4.1 본선에는 …개의 20'/40' 컨테이너를 적재할 것임.
.5 몇 입방미터의 화물 용적이 필요한가?
.5.1 …입방미터의 화물용적이 필요함.

.6 How many tonnes/cubic metres can the vessel still load?
.6.1 The vessel can still load … tonnes/cubic metres.
.7 How much deck cargo can the vessel load?
.7.1 The vessel can load … tonnes/cubic metres/… 20'/40' containers on deck.
.8 How many cars/trailers/trucks/… can the vessel load?
.8.1 The vessel can load … cars/trailers/trucks/… .
.9 What is the size of the hatch openings?
.9.1 The size of the hatch openings is … by … metres.
.10 What is the safety load of no. … hold?
.10.1 The safety load of the … deck of no. … hold is … tonnes per square metre.
.11 The vessel will still bunker … tonnes of fuel/fresh water/ ….

.2 Dockside/shipboard cargo handling gear and equipment

.1 Are dockside/floating cranes available?
.1.1 Yes, dockside/floating cranes are available.
.1.2 No, dockside/floating cranes are not available.
.2 What is the capacity of the crane?
.2.1 The capacity of the crane is … tonnes.
.3 What is the maximum reach of the crane?
.3.1 The maximum reach of the crane is … metres.
.4 What is the handling capacity of the container crane/ gantry?
.4.1 The handling capacity of container crane/gantry is … containers per hour.
.5 What is the handling capacity of the grain elevator/ore loader/…?
.5.1 The handling capacity of the grain elevator/ore loader/… is … tonnes/cubic metres per hour.
.6 What is the pumping capacity of the cargo pumps?
.6.1 The pumping capacity of the cargo pumps is …tonnes per hour.
.7 Are (light) fork-lift trucks for the cargo holds available?

.6 본선은 몇 톤/입방미터를 아직도 더 실을 수 있는가?
.6.1 본선은 …톤/입방미터를 아직도 더 실을 수 있음.
.7 본선은 갑판적 화물을 얼마나 적재할 수 있는가?
.7.1 본선은 갑판상에 …톤/입방미터/ …개의 20'/40' 컨테이너를 적재할 수 있음.
.8 본선은 얼마나 많은 자동차/트레일러/트럭/…을 적재할 수 있는가?
.8.1 본선은 …대의 자동차/트레일러/트럭/…을 적재할 수 있음.
.9 해치 개구의 규격은 얼마인가?
.9.1 해치 개구의 규격은 … x … 미터임.
.10 제 …번 화물창의 안전 하중은 얼마인가.
.10.1 제 …번 화물창의 …갑판의 안전 하중은 평방미터당 …톤임.
.11 본선은 …톤의 연료/청수/…을 아직도 더 적재할 것임.

.2 부두/선상의 하역 기기 및 설비

.1 부두/부선의 크레인은 사용 가능한가?
.1.1 예, 부두/부선의 크레인은 사용 가능함.
.1.2 아니오, 부두/부선의 크레인은 사용 가능하지 않음.
.2 크레인의 용량은 얼마인가?
.2.1 크레인의 용량은 …톤임.
.3 크레인의 최대 도달거리는 얼마인가?
.3.1 크레인의 최대 도달거리는 …미터임.
.4 컨테이너 크레인/갠트리의 하역능력은 얼마인가?
.4.1 컨테이너 크레인/갠트리의 하역능력은 시간당 …컨테이너임.
.5 곡물 하역기/광석 적재기/ …의 하역능력은 얼마인가?
.5.1 곡물 하역기/광석 적재기/ …의 하역능력은 시간당 …톤/입방 미터임.
.6 화물 펌프의 펌핑 용량은 얼마인가?
.6.1 화물 펌프의 펌핑 용량은 시간당 …톤임.
.7 화물창용(경량의) 지게차는 사용 가능한가?

.7.1 Yes, (light) fork-lift trucks are available.
.7.2 No, (light) fork-lift trucks are not available.
.8 Only use electric fork-lift trucks in the holds.
.9 What is the capacity of the fork-lift truck?
.9.1 The capacity of the fork-lift truck is … tonnes.
.10 What is the capacity of the derricks/cranes of the vessel?
.10.1 The capacity of the derricks/cranes of the vessel is … tonnes.
.11 What is the capacity of the … slings?
.11.1 The capacity of the … slings is … tonnes.
.12 These slings do not permit safe cargo handling.
.12.1 Replace the slings.
.13 Are bob-cats available for trimming?
.13.1 Yes, bob-cats are available for trimming.
.13.2 No, bob-cats are not available for trimming.

.3 Preparing for loading/unloading

.1 Prepare the vessel for loading/discharging.
.2 Unlock the hatch covers.
.3 Rig the hatchrails in no. … hold(s).
.4 Give notice of readiness to load/discharge by … UTC/localtime.
.5 Is the cargo list available and complete?
.5.1 Yes, the cargo list is available and complete.
.5.2 No, the cargo list is not available and complete (yet).
.5.3 The cargo list will be available and complete in … minutes.
.6 Complete the stowage plan.
.7 Make the stability calculation.
.8 Are the holds clean/dry/free of smell?
.8.1 Yes, the holds are clean/dry/free of smell.
.8.2 No, the holds are not clean/dry/free of smell (yet).
.8.3 The holds will be clean/dry/free of smell in … minutes/ hours.
.8.3.1 Clean the hold(s)/deck(s).

.7.1 예, (경량의) 지게차는 사용 가능함.
.7.2 아니오, (경량의) 지게차는 사용 가능하지 않음.
.8 화물창 내에서는 전기식 지게차만 사용하시오.
.9 지게차의 용량은 얼마인가?
.9.1 지게차의 용량은 …톤임.
.10 본선의 데리크/크레인의 용량은 얼마인가?
.10.1 본선의 데리크/크레인의 용량은 …톤임.
.11 …슬링의 용량은 얼마나 되는가?
.11.1 …슬링의 용량은 …톤임.
.12 이 슬링들은 안전 하역을 가능하게 하지 않음.
.12.1 슬링을 교체하시오.
.13 화물고르기를 위하여 보브캣이 사용 가능한가?
.13.1 예, 화물고르기를 위하여 보브캣이 사용 가능함.
.13.2 아니오, 화물고르기를 위하여 보브캣이 사용 가능하지 않음.

.3 적하/양하의 준비

.1 본선을 적하/양하에 대하여 준비하시오.
.2 해치 카바의 잠금 장치를 푸시오.
.3 제 … 번 화물창의 해치레일을 설치하시오.
.4 세계시/지방시 …시까지 적하/양하의 준비 완료 통지를 하시오.
.5 화물목록은 사용 가능하고 완전한가?
.5.1 예, 화물목록은 사용 가능하고 완전함
.5.2 아니오, 화물목록은 (아직)사용 가능하거나 완전하지 않음.
.5.3 화물목록은 …분 후 사용 가능하고 완전해짐.
.6 화물적부도를 완성하시오.
.7 복원성 계산을 하시오.
.8 화물창은 청결/건조/냄새 제거 되었는가?
.8.1 예, 화물창은 청결/건조/냄새 제거되었음.
.8.2 아니오, 화물창은 (아직) 청결/건조/냄새 제거 되지 않았음.
.8.3 화물창은 …분/시간 후 청결/건조/냄새제거 될 것임.
.8.3.1 화물창/갑판을 청소하시오.

.9 Are the safety arrangements in the hold(s) operational?
.9.1 Yes, the safety arrangements in the hold(s) are operational.
.9.2 No, the safety arrangements in the hold(s) are not operational(yet).
.9.3 The safety arrangements in the hold(s) will be operational in … minutes.
.10 Fill the double bottom tank(s)/ballast tank(s) before loading the heavy lifts.
.11 What is the maximum loading rate/discharging rate?
.11.1 The maximum loading rate/discharging rate is … tonnes per hour.
.11.2 Do not exceed the loading rate/discharging rate of … tonnes per hour.

.4 Operating cargo handling equipment and hatches

.1 Open all hatches before loading/discharging.
.2 Are the cranes/derricks operational?
.2.1 Yes, the cranes/derricks are operational.
.2.2 No, the cranes/derricks are not operational (yet).
.2.3 The cranes/derricks will be operational in … minutes.
.3 Rig the derrick(s)/crane(s) of no. … hold(s).
.4 Check the preventers.
.5 Keep within the safe working load of derrick(s)/crane(s).
.6 Instruct the winchmen/cranemen.
.7 Clean the tween deck(s) before opening lower hold(s).
.8 Switch on/off the hold ventilation.
.9 Switch on/off the hold lights.
.10 Close/open the cargo port(s) to no. … hold(s).

.5 Maintaining/repairing cargo handling equipment

.1 Check the hold(s)/hatch cover(s)/derrick(s)/… for damage and report.

.9 화물창 내의 안전 설비는 작동 가능한가?
.9.1 예, 화물창 내의 안전 설비는 작동 가능함.
.9.2 아니오, 화물창 내의 안전설비는 (아직)작동 가능하지 않음.
.9.3 화물창 내의 안전 설비는 …분 후 작동 가능할 것임.
.10 중량 화물을 적하하기 전에 이중저 탱크/밸러스트 탱크를 채우시오.
.11 최대의 적하/ 양하 속도는 얼마인가?
.11.1 최대의 적하/ 양하 속도는 시간당 …톤임.
.11.2 적하 속도/ 양하 속도가 시간당 ….톤을 초과하지 않도록 하시오.

.4 하역 설비와 해치의 작동

.1 적하/ 양하 이전에 모든 해치를 여시오.
.2 크레인/ 데리크는 작동 가능한가?
.2.1 예, 크레인/ 데리크는 작동 가능함.
.2.2 아니오, 크레인/ 데리크는(아직) 작동되지 않음.
.2.3 크레인/ 데리크는 …분 후 작동 가능할 것임.
.3 제 … 번 화물창의 데리크/ 크레인을 준비하시오.
.4 프리벤터를 점검하시오.
.5 데리크/ 크레인의 안전 작업 하중을 지키시오.
.6 윈치맨/ 크레인맨을 교육하시오.
.7 하부 화물창을 개방하기 전에 중갑판을 청소하시오.
.8 화물창 통풍 장치를 켜시오/ 끄시오.
.9 화물창 조명등을 켜시오/ 끄시오.
.10 제…번 화물창의 화물 하역용 개구를 닫으시오/ 여시오.

.5 하역 설비의 정비/수리

.1 화물창/해치 카바/데리크/ …의 손상 여부를 점검하고 보고하시오.

.1.1 The hold(s)/hatch cover(s)/derrick(s)/… is/are in order.
.1.2 The cargo battens are damaged.
.1.3 The rubber seals of the hatch cover(s) are damaged.
.1.4 The preventer(s) of no. … hold(s) is/are damaged.
.1.5 The (Container) lashings are damaged.
.1.6 … is/are damaged.
.1.6.1 Replace the damaged … .
.2 The hold ladder(s) is/are bent.
.2.1 Straighten the hold ladder(s).
.3 Are the hold ventilators operational?
.3.1 Yes, the hold ventilators are operational.
.3.2 No, the hold ventilators are not operational (yet).
.3.3 The hold ventilators will be operational in … minutes.
.4 Are the winch motors operational?
.4.1 Yes, the winch motors are operational.
.4.2 No, the winch motor of no. … derrick is not operational (yet).
.4.3 The winch motor of no. … derrick will be operational in …
minutes.
.5 Check the repair works personally.

.6 Briefing on stowing and securing

.1 Check the
- ~ careful and safe stowage.
- ~ complete unloading.
- ~ proper use of handling gear.
- ~ careful separation of different lots.

.2 Close the hatches in case of rain/snow/… .
.3 Refuse damaged/crushed/renailed/wet/torn/resewn/…boxes /cartons/cases/crates/bags/… .

.1.1 화물창/해치 카바/데리크/…는 정상임.
.1.2 선측 내장재가 손상되었음.
.1.3 해치 카바의 고무 틈새 막이가 손상되었음.
.1.4 제 …번 화물창의 프리벤터가 손상되었음.
.1.5 (컨테이너)고박 용구가 손상되었음.
.1.6 …이 손상되었음.
.1.6.1 손상된 …을 교체하시오.
.2 화물창 사다리가 굽어졌음.
.2.1 화물창 사다리를 곧게 펴시오.
.3 화물창 통풍기는 작동 가능한가?
.3.1 예, 화물창 통풍기는 작동 가능함.
.3.2 아니오, 화물창 통풍기는 (아직)작동되지 않음.
.3.3 화물창 통풍기는 …분 후 작동 가능할 것임.
.4 윈치 모터는 작동 가능한가?
.4.1 예, 윈치 모터는 작동 가능함.
.4.2 아니오, 제 …번 데리크의 윈치 모터는 (아직) 작동되지 않음.
.4.3 제 …번 데리크의 윈치 모터는 …분 후 작동 가능할 것임.
.5 수리 작업을 직접 점검하시오.

.6 적부와 고박에 관한 상황 설명

.1 적부가 주의 깊고 안전한가를 점검하시오.
양하가 완료되었는가 점검하시오.
하역 장비의 사용이 적절한가 점검하시오.
서로 다른 화물묶음이 주의 깊게 분리되었는지 점검하시오.
.2 강수/강설/…의 경우에 해치를 닫으시오.
.3 손상된/으깨진/다시 못질 된/젖은/찢어진/다시 꿰매진/…된 박스/판지 상자/케이스/나무틀/부대/…는 거절 하시오.

.4 Do not overstow cartons with other goods … .
.5 Do not use hooks for handling bags.
.6 Stow ventilation ducts into the bag cargo.
.7 Place dunnage between the tiers.
.8 Stow the
~ … into tween deck of no. … hold.
~ pallets/cartons/… closely together.
~ … in reefer hold.
~ empty containers in topmost tiers.
~ container(s) onto hatch cover(s).
~ … .
.9 Check the
~ containers for damage.
~ correct interlock of the stowpieces.[106)]
~ correct fixing of the rope clips.
~ … .
.10 Secure the heavy lift(s) immediately.
.11 Relash all lashings.

B3/1.2 Handling dangerous goods

Also see IMO-IMDG Code, London 1994, as revised.

.1 Briefing on nature of dangerous goods

.1 What is the IMO-Class of these goods?
.1.1 The IMO-Class of these goods is: … .
.2 This package contains IMO - Class … goods.
.3 These goods are flammable/poisonous/… .
.3.1 Handle these goods with caution.
.4 These goods emit flammable gases in contact with water.
.4.1 Keep these goods dry.
.5 These goods are liable to spontaneous heating and combustion.[107)]

.4 판지 상자 위에 다른 화물 …을 적부하지 마시오.
.5 부대를 취급하는데 갈고리를 사용하지 마시오.
.6 부대 화물 사이에 통풍관을 적부하시오.
.7 층 사이에 던네지를 놓으시오.
.8 제…번 화물창의 중갑판에 …을 적부하시오.
팔레트/판지 상자/…을 서로 밀착시켜 적부하시오.
…을 냉동 화물창에 적부하시오.
빈 컨테이너를 최상층에 적부하시오.
컨테이너를 해치 카바 위에 적부하시오.
…을 적부하시오.
.9 컨테이너가 손상되었는지를 점검하시오.
적부 용구가 서로 올바르게 맞물렸는지 점검하시오.
로프 클립이 올바르게 고정되었는지 점검하시오.
…을 점검하시오.
.10 중량 화물을 즉시 고박하시오.
.11 모든 고박용구를 다시 고박하시오.

B3/1.2 위험 화물의 취급.

IMO-IMDG 코드, London 1994도 참조

.1 위험 화물의 특성에 관한 상황 설명

.1 이들 화물의 IMO 등급은 무엇인가?
.1.1 이들 화물의 IMO 등급은: …임.
.2 이 꾸러미에는 IMO 등급 …의 화물이 들어있음.
.3 이들 화물은 가연성/유독성/…임,
.3.1 이들 화물을 주의하여 취급하시오.
.4 이들 화물은 물과 접촉하면 가연성 가스를 방출함.
.4.1 이들 화물을 건조하게 유지하시오.
.5 이들 화물은 자가 발열하고 연소하기가 쉬움.

.6 Do not touch … .

.2 Instructions on compatibility and stowage

.1 Observe the IMDG-Code when loading/stowing.
.2 Check the
~ proper segregation of goods.
~ correct technical names in documents.
~ correct marks/labels.
~ compatibility of IMO-Class … goods.
.3 Stow IMO-Class … goods
~ away from living quarters/away from … .
~ separated (by one hold) from IMO-Class … goods.
~ under/on deck.
.3.1 Cover IMO-Class … goods on deck with tarpaulins/ ….
.4 Stow
~ flammable goods away from the engine room bulkhead/ … .
~ infectious substances separated by one hold/ compartment from foodstuffs.
~ … drums away from IMO-Class … goods at a minimum of … metres.
.5 Brief the stevedores on the dangerous goods in number … hold(s).
.6 Refuse damaged/wet/… packings with dangerous goods.
.7 Ventilate the hold(s) before entering.
.8 Load/unload IMO-Class … goods first.
.9 No smoking during loading/unloading.

.3 Reporting incidents

.1 Sling(s) with bottles/drums/… of IMO-Class … goods were

.6 …에 접촉하지 마시오.

.2 격리 특성과 적부에 관련된 지시

.1 적하/적부시 IMDG 코드를 준수하시오.

.2 화물의 적절한 격리 상태를 점검하시오.
서류상에 전문명칭이 올바른지 점검하시오.
화표/꼬리표가 올바른지 점검하시오.
IMO 등급 …의 화물의 격리상태를 점검하시오.

.3 IMO 등급 …의 화물을 거주구역/…으로부터 떨어져서 적부 하시오.
IMO 등급 …의 화물을 IMO 등급 …의 화물로부터 (화물창 한 개를) 격리시켜 적부하시오.
IMO 등급 …의 화물을 갑판하/갑판상에 적부하시오.

.3.1 갑판상의 IMO 등급 …의 화물을 방수포/ …으로 덮으시오.

.4 가연성 화물을 기관실 격벽/ …으로부터 떨어져서 적부 하시오.
감염성의 물질을 식료품으로부터 화물창/구획 한개 떨어져서 적부하시오.
…드럼을 IMO 등급 …의 화물로부터 최소 …미터 떨어져서 적부하시오.

.5 하역 인부에게 제…번 화물창의 위험화물에 대하여 설명하시오.

.6 손상된/젖은/… 포장의 위험화물을 거절하시오.

.7 진입하기 전에 화물창을 통풍하시오.

.8 IMO 등급 …의 화물을 먼저 적하/양하하시오.

.9 적하/양하시 금연하시오.

.3 사고의 보고

.1 IMO 등급 …화물이 든 병/드럼/…을 담은 슬링이 갑판/

dropped on deck/into no. … hold/on pier … .
.1.1 Liquid/powder/gas is spilling.
.2 Several drums/barrels/tanks/… are deformed (and leaking).
.3 The … container with IMO-Class … goods is spilling out of the door.
.4 Spilling substances of IMO - Class … escaped into the sea/harbour water.
.4.1 Inform the pollution control.
.5 Temperature in locker/container/… with IMO-Class … goods is increasing (rapidly).
.6 Orange/red/… smoke is developing from IMO-Class … goods (on deck).
.7 Explosion in no. … hold.
.7.1 Damage to gas tank/container/… .
.8 Minor/major fire in number … hold.
.8.1 Fire extinguished.
.8.2 IMO-Class … goods re-ignited.
.8.3 Fire under control.
.8.4 Fire not under control (yet).
.8.4.1 Operate the general emergency alarm.
.8.4.2 Call the harbour fire brigade/… .
.9 Report injured persons/casualties.
.9.1 No person injured.
.9.2 Number of injured persons/casualties is … .

.4 Action in case of incidents

.1 Take actions according to the Emergency Plan.
.2 Turn the vessel out of the wind - the spilling gas/smoke is toxic.
.3 Put on protective clothing and breathing apparatus.
.4 Stop the spillage.
.5 Let the spillage evaporate.

제…번 화물창/…부두에 추락하였음.

.1.1 액체/분말/가스가 새고 있음.

.2 몇 개의 드럼/통/탱크/…이 변형됨(새고 있음).

.3 IMO 등급 …의 화물 …컨테이너가 문에서 새고 있음.

.4 IMO 등급 …의 새나간 물질이 바다/항만수역으로 빠져나감.

.4.1 오염 방제 담당자에게 알리시오.

.5 IMO 등급 …의 화물이 든 격납간/컨테이너/…내의 온도가 (급격히)상승하고 있음.

.6 (갑판상)IMO 등급 …의 화물에서 오렌지색/적색/…색 연기가 생기고 있음.

.7 제…번 하물창에 폭발이 있었음.

.7.1 가스 탱크/컨테이너/…에 손상이 있음.

.8 제…번 하물창에 작은/큰 화재가 있음.

.8.1 화재는 소화되었음.

.8.2 IMO 등급…의 화물이 재발화하였음.

.8.3 화재는 진압되고 있음.

.8.4 화재는 (아직) 진압되고 있지 않음.

.8.4.1 일반 비상 경보를 작동하시오.

.8.4.2 항만 소방대/…을 호출하시오.

.9 부상자/사상자를 보고하시오.

.9.1 부상자 없음.

.9.2 부상자/사상자의 수는 …임.

.4 사고 발생시의 조치

.1 비상 계획에 따라 행동을 취하시오.

.2 선박이 바람을 받지 않게 선회하시오-새는 가스/연기가 유독성임.

.3 보호복과 호흡구를 착용하시오.

.4 누출을 중지시키시오.

.5 누출물을 증발시키시오.

.6 Remove the spillage with synthetic scoops.
.6.1 Use absorbents for the spillage.
.6.2 Do not touch the spillage.
.7 Separate contaminated goods from other goods.
.8 Cover contaminated goods with tarpaulins/… .
.9 Only open the container/hold/locker/… when smoking is stopped.
.10 Cool down the container/… with water.
.11 Ventilate the hold(s) carefully.
.12 Close the hatch - operate the fire extinguishing system.
.13 Fight the fire from a great distance.
.14 Flood no. … hold(s).
.15 Rescue persons.
.15.1 Take injured persons/casualties to a safe area.
.15.2 Provide first aid to injured persons/casualties.
.15.3 Call the ambulance.
.16 Take off and dispose contaminated clothing.
.17 Alter course for the nearest port (-inform on radio).

B3/1.3 Handling liquid goods, bunkers and ballast-pollution prevention

.1 Preparing safety measures

.1 Plug the scuppers/drip-trays and report.[108)]
.1.1 All scuppers/drip-trays are plugged.
.2 Close the sea-valves/discharges and report.
.2.1 All sea-valves/discharges are closed.
.3 Stand by absorbent materials and report.
.3.1 Absorbent materials standing by.
.4 Stand by spill control gear and report.
.4.1 Spill control gear standing by.

.6 합성 재료 삽으로 누출물을 제거하시오.
.6.1 누출물에 대하여 흡착제를 사용하시오.
.6.2 누출물에 접촉하지 마시오.
.7 오염된 화물을 다른 화물과 격리하시오.
.8 방수포/…으로 오염된 화물을 덮으시오.
.9 연기 발생이 중지된 경우에만 컨테이너/ 화물창/ 격납간/…을 개방하시오.
.10 물로 컨테이너/…을 냉각하시오.
.11 화물창을 주의하여 통풍하시오.
.12 해치를 닫고 -소화 설비를 작동하시오.
.13 멀리서 화재를 진화하시오.
.14 제…번 화물창에 물을 채우시오.
.15 사람을 구조하시오.
.15.1 부상자/사상자를 안전한 구역으로 데려가시오.
.15.2 부상자/사상자에게 응급처치를 하시오.
.15.3 구급차를 호출하시오.
.16 오염된 의복을 벗고 처리하시오.
.17 가장 가까운 항구를 향하여 변침하시오.
(-무선으로 알리시오)

B3/1.3 액체 화물, 연료 및 밸러스트의 취급-오염 방지

.1 안전 조치의 준비

.1 갑판 배수구/누출유 받이통을 막고 보고하시오.
.1.1 모든 갑판 배수구/누출유 받이통을 막았음.
.2 해수 밸브/배출구를 닫고 보고하시오.
.2.1 모든 해수 밸브/배출구는 닫혔음.
.3 흡착 재료를 준비하고 보고하시오.
.3.1 흡착 재료를 준비하였음.
.4 누출 제어 기구를 준비하고 보고하시오.
.4.1 누출 제어 기구를 준비하였음.

.5 Stand by emergency fire pump/foam monitor/fire extinguishers and report.

.5.1 Emergency fire pump/foam monitor/fire extinguishers standing by.

.6 Fit bonding wire and report.[109)]

.6.1 Bonding wire is fitted.

.7 Maintain contact on VHF Channels … with the bunker barge/oil terminal.

.8 Is the oil pollution prevention plan available?

.8.1 Yes, the oil pollution prevention plan is available.

.8.2 No, the oil pollution prevention plan is not available (yet).

.8.3 The oil pollution prevention plan will be available in … minutes.

.9 Instruct the pumpman/… and report.

.9.1 Pumpman/… is instructed.

.2 Operating pumping equipment

including: phrases for communication with bunker barge /oil terminal

.1 What is the (maximum) loading rate/discharge rate?

.1.1 The (Maximum) loading rate/discharge rate is …tonnes per hour.

.2 Is the COW system/inert gas system operational?

.2.1 Yes, the COW system/inert gas system is operational.

.2.2 No, the COW system/inert gas system is not operational (yet).

.2.3 The COW system/inert gas system will be operational in… minutes.

.3 When will crude oil washing start?

.3.1 Crude oil washing will start in … minutes.

.4 Are your tanks inerted?

.4.1 Yes, my tanks are inerted.

.4.2 No, my tanks are not inerted (yet).

.4.3 My tanks will be inerted in … minutes.

.5 비상 소화 펌프/포말 방사기/소화기를 준비하고 보고하시오.
.5.1 비상 소화 펌프/포말 방사기/소화기를 준비하였음.
.6 본딩 와이어를 장착하고 보고하시오.
.6.1 본딩 와이어가 장착되었음.
.7 연료 부선/ 오일 터미널과 VHF 채널 …으로 교신을 유지하시오.
.8 유류 오염 방지에 대한 계획서가 이용 가능한가?
.8.1 예, 유류 오염 방지에 대한 계획서가 이용 가능함.
.8.2 아니오, 유류 오염 방지에 대한 계획서가 (아직) 이용 가능하지 않음.
.8.3 유류 오염 방지에 대한 계획서가 …분 후 이용 가능할 것임.
.9 펌프맨/…을 교육하고 보고하시오.
.9.1 펌프맨/…을 교육하였음.

.2 펌핑 설비의 작동

연료 부선/ 오일 터미널과의 교신을 위한 어귀를 포함

.1 (최대)적하/양하 속도는 얼마인가?
.1.1 (최대)적하/양하 속도는 시간 당 …톤임.
.2 COW 장치/불활성 가스 장치는 작동 가능한가?
.2.1 예, COW 장치/불활성 가스 장치는 작동 가능함.
.2.2 아니오, COW 장치/ 불활성 가스 장치는 (아직)작동되지 않음.
.2.3 COW 장치/불활성 가스 장치는 …분 후 작동 가능할 것임.
.3 원유 세정은 언제 시작되는가?
.3.1 원유 세정은 …분 후 시작될 것임.
.4 귀선의 탱크는 불활성화 되었는가?
.4.1 예, 본선의 탱크는 불활성화 되었음.
.4.2 아니오, 본선의 탱크는 (아직)불활성화 되지 않았음.
.4.3 본선의 탱크는 …분 후 불활성화 될 것임.

.5 What is the pressure in the inerted tanks?
.5.1 The pressure in the inerted tanks is … bar.
.6 What is the pumping pressure?
.6.1 The pumping pressure is … bar.
.7 Can we connect the loading arm?[110)]
.7.1 Yes, you can connect the loading arm.
.7.2 No, you cannot connect the loading arm (yet).
.7.3 Connect the loading arm in … minutes.
.8 Inform … minutes before loading/discharge will start/finish.
.8.1 Loading/discharge will start/finish in … minutes.
.9 What is the back pressure for stripping?
.9.1 The backpressure for stripping is… bars.
.10 Are the cargo hoses/booms connected?
.10.1 Yes, the cargo hoses/booms are connected.
.10.2 No, the cargo hoses/booms are not connected (yet).
.10.3 The cargo hoses/booms will be connected in … minutes.
.11 Are the cargo hoses/booms disconnected?
.11.1 Yes, the cargo hoses/booms are disconnected.
.11.2 No, the cargo hoses/booms are not disconnected (yet).
.11.3 The cargo hoses/booms will be disconnected in … minutes.
.12 Are you ready to load/discharge?
.12.1 Yes, I am ready to load/discharge.
.12.2 No, I am not ready to load/discharge (yet).
.12.3 I will be ready to load/discharge in … minutes.
.13 Keep a safe working pressure.
.14 Open the valve(s) and report.
.14.1 All full open aboard/ashore.
.15 Close the valve(s) and report.
.15.1 All full closed aboard/ashore.
.16 Start pumping (slowly).
.17 Are you pumping/receiving?

.5 불활성화 된 탱크의 압력은 얼마인가?
.5.1 불활성화 된 탱크의 압력은 …바아임.
.6 펌핑 압력은 얼마인가?
.6.1 펌핑 압력은 …바아임.
.7 이쪽에서는 로딩암을 연결해도 되는가?
.7.1 예, 그쪽에서는 로딩암을 연결해도 됨.
.7.2 아니오, 그쪽에서는 (아직)로딩암을 연결하면 안됨.
.7.3 …분 후 로딩암을 연결하시오.
.8 적하/ 양하가 개시/종료되기 …분 전에 통지하시오.
.8.1 적하/ 양하는 …분 후 개시/종료될 것임.
.9 스트리핑에 대한 배압은 얼마인가?
.9.1 스트리핑에 대한 배압은 …바아임
.10 카고 호스/ 부움이 연결되었는가?
.10.1 예, 카고 호스/ 부움이 연결되었음.
.10.2 아니오, 카고 호스/ 부움이 (아직) 연결되지 않았음.
.10.3 카고 호스/ 부움은 …분 후 연결될 것임.
.11 카고 호스/ 부움은 분리되었는가?
.11.1 예, 카고 호스/ 부움은 분리되었음.
.11.2 아니오, 카고 호스/ 부움은 (아직) 분리되지 않았음.
.11.3 카고 호스/ 부움은 …분 후 분리될 것임.
.12 귀선은 적하/ 양하 준비되었는가?
.12.1 예, 본선은 적하/ 양하 준비되었음.
.12.2 아니오, 본선은 (아직) 적하/ 양하 준비되지 않았음.
.12.3 본선은…분 후 적하/ 양하 준비될 것임.
.13 안전 작업 압력을 지키시오.
.14 밸브를 열고 보고하시오.
.14.1 본선/ 육상 측을 완전히 열었음.
.15 밸브를 닫고 보고하시오.
.15.1 본선/ 육상 측을 완전히 닫았음.
.16 (천천히)펌핑을 시작하시오.
.17 그 쪽은 펌핑하고/받아들이고 있는가?

.17.1 Yes, I am pumping/receiving.
.17.2 No, I am not pumping/not receiving.
.18 Increase/decrease pumping rate to … revolutions/bar.
.19 (Quantity received) - stop pumping.

.3 Reporting and cleaning up spillage

.1 Leak at manifold connection!
.1.1 Overflow at … !
.2 Stop pumping !
.3 How much is spilled?
.3.1 Spill is about … tonne(s).
.4 Treat spill with … .
.5 Stand by oil clearance team and report.
.5.1 Oil clearance team standing by.
.5.2 All crew assist to remove the spill.
.5.2.1 Spillage stopped.
.5.2.2 Spill cleaned up.
.5.2.3 Spill waste contained in save-all/… .
.6 Oil/… escaping into sea/harbour water!
.6.1 Inform pollution control!

.4 Ballast handling

.1 Plug the scuppers and report.
.1.1 All scuppers are plugged.
.2 Open/close the sea suction valve/ballast tank valve no. … and report.
.2.1 Sea suction valve/ballast tank valve no. … is opened/closed.
.3 Start the ballast pump and report.
.3.1 Ballast pump started.
.4 Stop the ballast pump (- ballast overflow!) and report.
.4.1 Ballast pump stopped.

.17.1 예, 이 쪽은 펌핑하고/받아들이고 있음.
.17.2 아니오, 이 쪽은 펌핑하고/받아들이고 있지 않음.
.18 펌핑 속도를 …회전/바아까지 올리시오/내리시오.
.19 (수령량)-펌핑을 중지하시오.

.3 누출 관련 보고와 누출 제거

.1 매니폴드 연결부에서 새고 있음!
.1.1 …에서 넘치고 있음!
.2 펌핑을 중지하시오!
.3 얼마나 누출되었는가?
.3.1 누출량은 약…톤임.
.4 …으로 누출유를 청소하시오.
.5 유류 제거반을 대기하고 보고하시오.
.5.1 유류 제거반을 대기하고 있음.
.5.2 모든 승무원은 누출유 제거를 도우시오.
.5.2.1 누출이 중단되었음.
.5.2.2 누출이 청소되었음.
.5.2.3 누출용 걸레는 기름받이통/…에 들어 있음.
.6 유류/…가 해수/항만 수역으로 새고 있음!
.6.1 오염 제어 담당자에게 통지하시오.

.4 밸러스트 취급

.1 갑판 배수구를 막고 보고하시오.
.1.1 모든 갑판 배수구를 막았음.
.2 제…번 해수 흡입 밸브/밸러스트 탱크 밸브를 열고/닫고 보고 하시오.
.2.1 제…번 해수 흡입밸브/밸러스트 탱크 밸브는 열렸음/닫혔음.
.3 밸러스트 펌프를 시동하고 보고하시오.
.3.1 밸러스트 펌프를 시동하였음.
.4 밸러스트 펌프를 정지하고 보고하시오.(밸러스트가 넘치고 있음!)
.4.1 밸러스트 펌프를 정지하였음.

.5 Pump out ballast tank no. … and report.
.5.1 Ballast tank no. … is pumped out.
.5.2 Stop the ballast pump - ballast dirty!

.5 Cleaning tanks

.1 Pump the slops into the slop tank.[111)]
.2 Dispose the sludge into the sludge tank.
.3 Order a shore slop tank/slop barge.
.3.1 We have … tonnes of slops/sludge.
.4 Start/stop pumping slops.
.5 Keep a safe working pressure.

B3/1.4 Preparing for sea

.1 Close and secure the hatch covers for sea and report.
.1.1 Hatch covers closed and secured.
.2 Lash and secure the goods for sea and report.
.2.1 Goods lashed and secured.
(In ro/ro-ferries: The execution of the instructions 3,4 and 5 given from the bridge on radio should be confirmed from the person in charge of the corresponding station using phrases 3.1, 4.1 and 5.1)
.3 Close and secure the bow door/stern door and report.
.3.1 Bow door/stern door closed and secured.
.4 Fold and secure the bow ramp/stern ramp/side ramp and report.
.4.1 Bow ramp/stern ramp/side ramp folded and secured.
.5 Lash and secure all cars/trucks/wagons/… and report.
.5.1 All cars/trucks/wagons/… lashed and secured.
.6 Lower and secure the derricks/cranes and report.
.6.1 Derricks/cranes lowered and secured.
.7 Check the seaworthiness of the holds and report.

.5 제…번 밸러스트 탱크를 배출하고 보고하시오.
.5.1 제…번 밸러스트 탱크는 배출되었음.
.5.2 밸러스트 펌프를 정지하시오-밸러스트가 불결함!

.5 탱크 청소

.1 펌프로 슬롭을 슬롭 탱크에 주입하시오.
.2 슬러지(침전물)를 슬러지 탱크로 옮기시오.
.3 육상의 슬롭 탱크/슬롭 바아지를 요청하시오.
.3.1 본선은 …톤의 슬롭/슬러지 보유하고 있음
.4 슬롭의 펌핑을 개시/중지하시오.
.5 안전 작업 압력을 지키시오.

B3/1.4 출항 준비

.1 항해에 대비하여 해치카바를 닫고 고박하여 보고하시오.
.1.1 해치 카바를 닫고 고박하였음.
.2 항해에 대비하여 화물을 고박하고 보고하시오.
.2.1 화물을 고박하여 안전하게 되었음.
(ro/ro 연락선 내에서 : 선교로부터 무선으로 주어진 3, 4, 5항 지시의 수행 내용은 해당 부서의 책임자가 3.1, 4.1, 5.1의 어귀를 사용하여 확인해 주어야한다.)
.3 선수 출입문/선미 출입문을 닫아 안전하게 하고 보고하시오.
.3.1 선수 출입문/선미 출입문을 닫아 안전하게 하였음.
.4 선수램프/ 선미램프/ 선측램프를 접어 올려 안전하게 하고 보고하시오.
.4.1 선수램프/ 선미램프/선측램프를 접어 올려 안전하게 하였음.
.5 모든 자동차/ 트럭/ 왜건/ …을 고박하여 안전하게 하고 보고하시오.
.5.1 모든 자동차/트럭/왜건/…을 고박하여 안전하게 하였음.
.6 데리크/크레인을 내려 안전하게 하고 보고하시오.
.6.1 데리크/크레인을 내려 안전하게 하였음.
.7 화물창의 감항성을 점검하고 보고하시오.

.7.1 Holds seaworthy.
.8 How much ballast can we take (down to her marks)?
.8.1 We can take … tonnes of ballast.
.9 Check the trim.
.9.1 Fill the forepeak to decrease the stern trim.
.9.2 Fill the double-bottom tank(s).
.9.3 Pump fuel from … tank to … tank to bring the vessel upright.

B3/2 Cargo Care

B3/2.1 Operating shipboard equipment for cargo care

.1 Is the equipment for cargo care operational?
.1.1 Yes, the equipment for cargo care is operational.
.1.2 No, the …*(equipment)* is not operational (yet).
.1.3 The …*(equipment)* will be operational in … minutes.
.2 What is the air change rate of the hold ventilators?
.2.1 The air change rate of the hold ventilators is … -fold.
.3 Are the temperature/humidity recorders in the hold(s) operational?
.3.1 Yes, the temperature/humidity recorders in the hold(s) are operational.
.3.2 No, the temperature/humidity recorders in the hold(s) are not operational (yet).
.3.3 The temperature/humidity recorders in the hold(s)will be operational in … minutes.
.4 Instruct the crew how to connect reefer plugs/clip-on units/… and report.[112)]
.4.1 The crew is instructed how to connect reefer plugs /clip-on units/… .

.7.1 화물창은 감항성이 있음.
.8 본선은 (흘수선까지)밸러스트를 얼마나 적재할 수 있는가?
.8.1 본선은 밸러스트를 …톤 적재할 수 있음.
.9 트림을 점검하시오.
.9.1 선미 트림을 감소시키기 위하여 선수 탱크를 채우시오.
.9.2 이중저 탱크를 채우시오.
.9.3 본선을 똑바로 하기 위하여 연료유를 …탱크에서 …탱크로 펌프로 옮기시오.

B3/2 화물 관리

B3/2.1 본선의 화물 관리용 설비의 작동

.1 화물 관리용 설비는 작동 가능한가?
.1.1 예, 화물 관리용 설비는 작동 가능함.
.1.2 아니오, …(설비)는 (아직) 작동되지 않음.
.1.3 …(설비)는 …분 후 작동 가능할 것임.
.2 화물창 통풍기의 환기율은 얼마인가?
.2.1 화물창 통풍기의 환기율은 …배임.
.3 화물창 내의 온도/습도 기록기는 작동 가능한가?
.3.1 예, 화물창 내의 온도/습도 기록기는 작동 가능함.
.3.2 아니오, 화물창 내의 온도/습도 기록기는 (아직)작동되지 않음.
.3.3 화물창 내의 온도/습도 기록기는 …분 후 작동 가능할 것임.
.4 승무원들에게 냉동 플러그/클립온 유니트/…의 연결 방법을 교육하고 보고하시오.
.4.1 승무원들은 냉동 플러그/클립온 유니트/…의 연결방법을 교육 받았음.

B3/2.2 Taking measures for cargo care

.1 Carrying out inspections

.1 The holds must be inspected by the surveyor before loading.
.2 Check the reefer holds for proper loading preparation and report.
.2.1 The reefer holds are ready for loading.
.2.2 The reefer holds are not ready for loading (yet).
.2.3 The reefer holds will be ready for loading in … minutes.
.3 Are the holds clean (dry and free of smell)?
.3.1 Yes, the holds are clean(dry and free of smell).
.3.2 No, the holds are not clean (dry and free of smell) (yet).
.3.3 The holds will be clean (dry and free of smell) in … minutes/ hours.
.4 Check the operation of the hold ventilators and report.
.4.1 The hold ventilators are operational.
.4.2 The hold ventilators (in no. … hold(s)) are not operational (yet).
.4.3 The hold ventilators (in no … hold(s)) will be operational in … minutes.
.5 Order a surveyor to check the reefer plugs/cargo securings.
.6 Is the certificate of survey available and complete?
.6.1 Yes, the certificate of survey is available and complete.
.6.2 No, the certificate of survey is not available and complete (yet).
.6.3 The certificate of survey will be available and complete in … minutes/hours.
.7 Check the lashings and securings every day/… hours.
.8 Enter all checks into the log-book.[113)]
.9 Before unloading open the hatches only when the surveyor is present.

B3/2.2 화물 관리 조치의 실시

.1 검사의 실시

.1 화물창은 적하 전에 검사관에 의하여 검사되어야 함.
.2 냉동 화물창이 적절하게 적하 준비되었는지 점검하고 보고하시오.
.2.1 냉동 화물창은 적하 준비되었음.
.2.2 냉동 화물창은 (아직)적하 준비되지 않았음.
.2.3 냉동 화물창은 …분 후 적하 준비될 것임.
.3 화물창은 (건조하고 냄새가 없이) 청결한가?
.3.1 예, 화물창은 (건조하고 냄새가 없이) 청결함.
.3.2 아니오, 화물창은 (아직) (건조하고 냄새가 없이)청결하지 않음.
.3.3 화물창은 …분/시간 후 (건조하고 냄새가 없이)청결하게 될 것임.
.4 화물창 통풍기의 작동 상태를 점검하고 보고하시오.
.4.1 화물창 통풍기는 작동 가능함.
.4.2 (제…번 화물창의) 화물창 통풍기는 (아직)작동되지 않음.
.4.3 (제…번 화물창의) 화물창 통풍기는 …분 후 작동 가능함.
.5 검사관에게 냉동 플러그/화물 고정 상태를 점검하도록 지시하시오.
.6 검사 증명서가 유효하고 완전한가?
.6.1 예, 검사 증명서가 유효하고 완전함.
.6.2 아니오, 검사 증명서가 (아직) 유효하고 완전하지 못함.
.6.3 검사 증명서는 …분/시간 후에 유효하고 완전하게 될 것임.
.7 고박과 고정 상태를 매일/매…시간마다 점검하시오.
.8 모든 점검 사항을 항해일지에 기재하시오.
.9 양하전에 오로지 검사관의 참석 시에만 해치를 개방하시오.

.2 Describing damage to the cargo

Also see section B2/1.2.3 "Reporting incidents"

.1 The …*(cargo)* is in bad condition.

.2 The packages of …*(cargo)* are
~ wet/damp/mouldy.
~ marked by fresh water/sea water.

.3 The metal of …*(cargo)* is rusty.

.4 The bands of …*(cargo)* are broken/missing/rusty.

.5 The crates/cases with …*(cargo)* are renailed.

.5.1 The boards of crates/cases with …*(cargo)* are loose.

.6 The marks/labels on …*(cargo)* are unclear/illegible/ false.

.7 The contents of drums/barrels/… are unknown.

.8 The weight of the …*(cargo)* is unknown.

.9 The boxes/crates/cases/… with …*(cargo)* are damaged.

.10 The bags/bales with …*(cargo)* are torn/resewn/spilling.

.11 The drums/barrels/… with …*(cargo)* are deformed/spilling.

.12 The boxes/cartons/cases/… with …*(cargo)* are crushed.

.13 The bags/boxes/cartons/… with …*(cargo)* are not full/slack/ empty.

.14 The bags/boxes/cartons/… with …*(cargo)* are second hand.

.15 The boxes/cartons/cases/… with bottles of …*(cargo)* are (partly)broken.

.16 The …*(cargo)* is (partly)
~ eaten by rats /worms.
~ infected with vermin.
~ missing.

.17 … container(s) are damaged.

.17.1 … container(s) were damaged
~ before loading.
~ during loading.
~ by shifting on board.
~ by heavy seas.

.2 화물 사고의 설명

(B2/ 1.2.3 "사고의 보고" 부분도 참조)

.1 …(화물)이 나쁜 상태임.
.2 …(화물)의 포장재가 젖었음/습기 찼음/곰팡이 났음.
…(화물)의 포장재가 청수/해수로 자국이 났음.
.3 …(화물)의 금속 부분이 녹슬었음.
.4 …(화물)의 밴드가 부서졌음/없어졌음/녹슬었음.
.5 …(화물)의 나무틀/상자가 다시 못질되었음.
.5.1 …(화물)의 나무틀/상자의 판자가 느슨함.
.6 …(화물)의 화표/꼬리표가 불명확함/판독하기 어려움/틀렸음.
.7 드럼/통/…의 내용물을 알 수 없음.
.8 …(화물)의 중량을 알 수 없음.
.9 …(화물)의 박스/나무틀/상자/…이 손상되었음.
.10 …(화물)의 부대/꾸러미가 찢어졌음/다시 꿰맸음/새고 있음.
.11 …(화물)의 드럼/통/…이 변형되었음/새고 있음.
.12 …(화물)의 박스/판지 상자/상자/…이 으깨졌음.
.13 …(화물)의 부대/박스/판지 상자/…이 꽉 차지 않았음/헐거움/ 비었음.
.14 …(화물)의 부대/박스/판지 상자/…가 중고임.
.15 …(화물)이든 병의 박스/판지 상자/상자/…이 (부분적으로) 부서졌음.
.16 …(화물)이 (부분적으로) 쥐/벌레에 먹혔음.
…(화물)이 (부분적으로) 해충에 감염되었음.
…(화물)이 (부분적으로) 없어졌음.
.17 …컨테이너가 손상되었음.
.17.1 …컨테이너가 선적 전에 손상되었음.
…컨테이너가 선적 중에 손상되었음.
…컨테이너가 선상에서 이동되다가 손상되었음.
…컨테이너가 큰 파도에 의하여 손상되었음.

.18 … container(s) were washed overboard (-inform on radio).
.19 The temperature in no. … hold is above normal/below normal/critical/… degrees Celsius.
.20 The humidity of …*(cargo)* is above normal/below normal /critical.

.3 Taking actions

Also see section B2/1.2.4 "Action in case of incident"

.1 Switch on the hold ventilation to supply/exhaust air.
.2 Switch off the hold ventilation (in case of shipping seas).
.3 Switch on/off the automatic temperature control/recorder.
.4 Relash the container(s)/car(s)/trucks(s)… in no. … hold/on … deck.
.5 Replug the reefer container(s) in no. … hold/on deck.
.6 Secure the shifting cargo in no. … hold/on … deck.
.7 Protect the deck cargo of …*(cargo)* against sun/rain shipping seas.
.8 Keep the deck cargo of …*(cargo)* wet/dry.
.9 Check the contents of drum(s)/barrel(s)/container(s)/… with false labels.

.18 …컨테이너가 선외로 쓸려갔음.(-무선으로 통지하시오)

.19 제…번 화물창 내의 온도가 정상치 보다 높음/정상치 보다 낮음 / 임계치 임/ 섭씨…도임.

.20 …(화물)의 습도가 정상치 보다 높음/정상치 보다 낮음/임계치임.

.3 조치 의 실행

(B2/1,2,4 "사고 발생 시의 조치" 참조)

.1 공기를 공급/배출하기 위하여 화물창 통풍기를 켜시오.

.2 (파도가 올라올 경우)화물창 통풍기를 끄시오.

.3 자동 온도 조절/기록 장치를 켜시오/끄시오.

.4 제…번 화물창/…갑판상의 컨테이너/자동차/트럭/…을 다시 고박하시오.

.5 제…번 화물창 내의/갑판상의 냉동컨테이너의 플러그를 다시 끼우시오.

.6 제…번 화물창/…갑판상의 이동 화물을 고정하시오.

.7 갑판적 화물 …(화물)을 햇빛/강수/파도로부터 보호하시오.

.8 갑판적 화물 …(화물)을 축축하게/건조하게 유지하시오.

.9 꼬리표가 잘못된 드럼/통/컨테이너/…의 내용물 점검하시오.

B4 Passenger Care

The phrases of this chapter should help Masters, officers and crew members of passenger vessels and passenger ferries to inform passengers on safety aspects and to manage them in case of an emergency.

B4/1 Briefing and Instruction

B4/1.1 Conduct of passengers on board

.1 General information on conduct of passengers

.1 Ladies and Gentlemen. This is captain … speaking.
I have pleasure in informing you that all safety equipment is in full working order. The bow/stern doors are closed and secured. The vessel is in all respects ready for sea.
Please listen carefully to the safety instructions which follow. In the unlikely event of an emergency, please obey the orders given on the public address system.

.2 Passengers are requested to read all notes and leaflets concerning safety regulations.

.3 All regulations concerning the vessel's routine have to be obeyed.

.2 Briefing on prohibited areas, decks, and spaces

Safety regulations do not permit passengers to enter the following spaces:

- navigating bridge
- engine room
- manoeuvring areas at the front and back end of the vessel

B4 여객의 관리

이장의 어귀들은 여객선과 여객용 연락선에서 여객에게 안전 관련 통보를 하거나, 비상시에 여객을 관리하기 위하여 선장, 항해사, 승무원들에게 도움이 될 것임.

B4/1 상황 설명 및 교육

B4/1.1 선상에서의 여객의 안내

.1 여객의 안내에 관한 일반 사항

.1 신사, 숙녀 여러분. 선장…가 말씀드립니다.
모든 안전 설비는 완전한 작동 상태에 있다는 것을 말씀드리게 됨을 기쁘게 생각합니다. 선수/선미 출입문은 닫혀 안전하게 되었습니다. 본선은 모든 측면에서 항해 준비가 되었습니다. 다음의 안전 지시 사항을 주의 깊게 경청해 주시기 바랍니다.
있음직하지는 않으나 비상사태가 발생하면, 확성방송장치로 하달되는 지시에 따르시기 바랍니다.

.2 여객들은 안전 규정에 관련된 모든 주의 사항과 전단 인쇄물을 읽어야 합니다.

.3 선박의 통상 업무에 관련된 모든 규정은 준수되어야 합니다.

.2 출입 금지 구역, 갑판, 장소에 대한 설명

안전규정에 의하여 여객들은 다음 장소에 들어가서는 안됨:
- 항해 선교
- 기관실
- 선박의 선수, 선미 끝 부분에 있는 조종용의 구역

- cargo rooms and compartments
- service rooms
- all areas and spaces marked "Crew only"
- all closed, sealed or roped off areas, spaces and rooms
- car decks when the vessel is at sea.

B4/1.2 Briefing on safety regulations, preventive measures and communications

.1 Drills[114)]

.1 International regulations require all passengers to be assembled in a drill which has to take place within 24 hours of departure.

.2 A drill will be held to familiarize passengers with their assembly stations, with their life-saving equipment and with emergency procedures.

.3 All passengers must attend this drill.

.2 The general emergency alarm

.1 In case of emergency seven short blasts and one prolonged blast will be given with the ship's whistle and the alarm system.

.2 Remain calm when you hear the general emergency alarm.

.3 Passengers will be taught how to act and behave in cases of emergency.

.3 Preventing/reporting fire

.1 Always remember that fire is the greatest hazard aboard ship.

.2 Always act immediately if you detect fire, smell, fumes or smoke.

.3 Always inform a member of the crew if you detect fire, smell fumes or smoke.

- 화물실과 구획
- 운항실
- "승무원에게만"이라는 표시가 되어 있는 구역과 장소
- 모든 밀폐, 봉인 또는 로프로 격리된 구역, 장소와 방
- 선박이 외해에 있을 경우의 차량 갑판

B4/1.2 안전 규정, 예방 조치, 의사소통에 관한 설명

.1 훈련

.1 국제 규정에 의하여 모든 여객은 출항 후 24시간 이내에 실시되는 훈련에 참가하여야 함.

.2 훈련은 여객으로 하여금 그들의 집합 장소를 숙지하도록 하기 위하여, 그들의 구명 설비와 비상 절차를 숙지하도록 하기 위하여 실시될 것임.

.3 모든 여객은 이러한 훈련에 참가하여야 함.

.2 일반 비상경보

.1 비상사태 발생시 선박의 기적과 경보 장치로 단성 7발과 장성 1발이 울릴 것임.

.2 일반 비상경보를 들었을 때 조용히 있어야 함.

.3 여객들은 비상사태시 어떻게 행동할 것인가에 대하여 교육을 받을 것임.

.3 화재의 예방/보고

.1 선상에서 화재는 가장 큰 위험요소라는 것을 항상기억할 것.

.2 불, 냄새, 유해 가스 또는 연기를 감지하면 항상 즉각적으로 행동할 것

.3 불, 냄새, 유해 가스 또는 연기를 감지했을 때 항상 승무원 중 한 사람에게 알릴 것.

.4 Be careful to extinguish cigarettes completely.
.5 Put used cigarettes in a container provided.
.6 Never smoke in bed.
.7 Never smoke on deck except in areas labelled as smoking areas.
.8 Never throw a cigarette overboard.
.9 The use of naked light and open fire is strictly prohibited.
.10 Never use lighted candles.
.11 Never hang anything over or near an electric bulb.
.12 Never use an electric iron in a cabin.
If you need to iron something use the ironing room on … deck.
The key may be collected at the information desk.
.13 If you detect a fire, smell, fumes or smoke act immediately as follows:
- Call out "Fire!"
- Operate the nearest fire alarm
- Inform a member of the crew
- Telephone the navigating bridge. The number to dial is ….

.4 PA announcements on emergency[115)]

.1 Attention please! Attention please!
This is your captain with an important announcement.
I repeat, this is your captain with an important announcement.
.1.1 We have grounded/a minor flooding (in..)/a minor fire (in..).
.1.2 There is no immediate danger to our passengers or the ship
- and there is no reason to be alarmed.
.1.3 For safety reasons we request all passengers to go to their assembly stations on …deck and wait there for further instructions.
.1.4 Please follow the instructions given by the officers and crew.
.1.5 The ship's fire fighting team/damage control team is fighting the fire/flooding.

.4 담배 불을 완전히 끄도록 주의할 것.
.5 꽁초는 마련된 용기에 버릴 것.
.6 침대에서는 절대 흡연하지 말 것.
.7 흡연 구역이라고 표시된 곳 이외의 갑판에서는 절대 흡연하지 말 것.
.8 꽁초를 선외로 버리지 말 것.
.9 나화와 노출된 불꽃을 사용하는 것은 엄격하게 금지됨.
.10 촛불을 절대 사용하지 말 것.
.11 전구 위나 근처에 물건을 절대 매달지 말 것.
.12 선실 내에서 전기다리미를 절대 사용하지 말 것.
무엇인가를 다림질할 필요가 있으면 …갑판에 있는 다림질 방을 이용할 것. 열쇠는 안내소에서 받을 수 있음.
.13 만약 불, 냄새, 유해 가스 또는 연기를 감지했을 때는 즉시 다음과 같이 행동할 것.
- "불이야!" 라고 외칠 것.
- 가장 가까운 화재경보기를 작동시킬 것.
- 승무원 중 한 사람에게 알릴 것.
- 항해 선교에 전화할 것. 전화번호는 …임.

.4 비상시의 확성 장치 방송

.1 주목하여 주시오! 주목하여 주시오!
본인은 선장이며 중요한 공지사항을 알립니다. 반복합니다, 본인은 선장이며 중요한 공지사항을 알립니다.
.1.1 본선은 좌초/(…에)경미한 침수/(…에)경미한 화재가 발생함.
.1.2 여객이나 본선에 즉각적인 위험은 없습니다.
- 따라서 경보를 발할 이유는 없습니다.
.1.3 안전상의 이유로, 모든 승객은 …갑판에 있는 각자의 집합 장소로 가, 거기서 추가의 지시를 기다리기 바랍니다.
.1.4 사관과 승무원이 내리는 지시에 따르기 바랍니다.
.1.5 본선의 소화반/손상 제어반이 소화작업/방수작업을 하고 있습니다.

.1.6 We also have radio contact with other ships/coast radio stations.

.1.7 The fire/flooding is under control.

.1.8 As soon as I have further information I will make another announcement. I ask you kindly to remain calm. There is no danger at this time.

.2 This is your Captain speaking. I have another announcement. The fire/flooding is not under control yet.

.2.1 There is smoke/flooding in … - access to this area is prohibited.

.2.2 For safety reasons we request all passengers to prepare to go to their assembly stations. Access to the assembly stations will be via …. Do not forget to take your life-jackets and blankets with you.

.2.3 All passengers of deck no. … are requested to follow the crew members who will escort you to your assembly stations.

.2.4 When you get to your assembly stations put on your life-jackets and wait for further orders.

.2.5 Do not go to the lifeboat stations until you are ordered to do so.

.2.6 Go to your lifeboat stations.

.2.7 Follow the escape routes shown.

.2.8 Do not enter the lifeboats/liferafts. The order to enter the life-boats/liferafts will be given from the bridge or by the officers.

.2.9 We have just received a message from shore/other vessels that assistance is on the way.
Assistance should arrive within approximately … hours.

.5 Person overboard

.1 If you see anybody fall overboard, act as follows:
- call out "Man overboard“

.1.6 본선은 타선/해안 무선국과도 무선 교신을 하고 있습니다.
.1.7 화재/침수는 진압되고 있습니다.
.1.8 추가의 통보 사항이 있으면 즉시 다시 방송하겠습니다.
본인은 여러분이 조용히 있어주시기를 요청합니다.
현재로서는 위험이 없습니다.
.2 선장이 전합니다. 추가의 방송을 하겠습니다.
화재/침수는 아직 진압되지 않고 있습니다.
.2.1 …에 연기가/침수가 있습니다.
-이 구역에는 접근 금지입니다.
.2.2 안전상의 이유로, 모든 승객은 각자의 집합 장소로 갈 준비를 하시기 바랍니다. 집합 장소로의 접근은 …을 통하여 이루어질 것입니다.
구명동의와 담요를 잊지 말고 지참하시기 바랍니다.
.2.3 제…번 갑판의 모든 승객은 승무원을 따라가시오, 그러면 승무원이 각자의 집합 장소로 인도할 것입니다.
.2.4 각자의 집합 장소에 도착하면 구명동의를 착용하고 추가의 지시를 기다리시오.
.2.5 그렇게 하도록 지시될 때까지는 구명정 부서에 가지 마시오.
.2.6 각자의 구명정 부서로 가시오.
.2.7 안내된 탈출 통로를 따라가시오.
.2.8 구명정/구명벌에 들어가지 마시오. 구명정/구명벌에 들어가라는 명령은 선교 또는 사관으로부터 내려질 것입니다.
.2.9 우리는 육상/타선박으로부터 원조대가 오고 있다는 통보를 수신하였음.
원조대는 대략… 시경까지 도착할 것임.

.5 사람의 선외 추락

.1 누군가 선외로 떨어진 것을 보았을 때, 다음과 같이 행동하시오:
- "사람이 선외로 떨어졌다"라고 외치시오.

- throw lifebuoys overboard
- keep your eyes on the person in the water
- show/tell an officer/crew the person's position in the water, or telephone the bridge immediately, the number is ….

.6 Protective measures for children

.1 Children must be kept under permanent observation.

.2 Never let children climb or sit on the ship's rails.

.3 Special lifejackets for children are available; please ask the steward/stewardess.

.4 You may leave your children under qualified care in the children's playroom/on the playdeck on … deck from … to … hours.

B4/2 Evacuation and Boat Drill

B4/2.1 Allocating/directing to assembly stations, describing how to escape

.1 When the general emergency alarm is sounded which consists of seven short blasts and one prolonged blast, all passengers have to go to their assembly station. Take your lifejackets and blankets with you. Life jackets are stored in your cabins under your beds and at your assembly stations.

You are encouraged to try on your lifejacket.

.2 All passengers must put on.

- warm clothing.
- long trousers, long-sleeved shirts/jackets.
- strong shoes and head covering.

.3 All passengers with their lifejackets and blankets are requested to go to their assembly stations/the lounge/th e… immediately.

- 구명부환을 선외로 던지시오.
- 시선을 물에 있는 사람에게 고정하시오.
- 물에 빠진 사람의 위치를 사관/승무원에게 보이시오/ 말하시오 또는 즉시 선교에 전화하시오.
  전화번호는 …번임.

.6 어린이에 대한 보호 조치

.1 어린이는 항상 관찰 하에 두어야함.

.2 어린이로 하여금 선박의 난간에 기어오르거나 앉아 있게 하여서는 절대 안됨.

.3 어린이용의 특별 구명동의가 사용 가능함; 남/녀 여객계원에게 요청하시오.

.4 어린이는 …갑판의 어린이 놀이방/놀이갑판에서 …시에서 … 시까지 유자격자가 보살피도록 하여도 됨.

B4/2 대피 및 단정 훈련

B4/2.1 집합 장소 할당/집합 장소로의 인도, 탈출방법에 대한 설명

.1 단성 7발과 장성 1발로 구성되는 일반 비상경보가 울리면, 모든 여객은 각자의 집합 장소로 가야함. 각자의 구명동의와 담요를 지참하시오. 구명동의는 각자 선실의 침대 밑과 각자의 집합 장소에 있음.
각자는 각자의 구명동의를 입어 보기를 권고함.

.2 모든 여객은 다음을 착용해야함.
-보온용 의복
-긴 바지, 긴소매 셔츠/웃옷
-단단한 신발과 머리 보호구

.3 각자의 구명동의와 담요를 지참한 모든 여객은 각자의 집합 장소/ 휴게실/ …에 즉시 가시오.

.4 From your assembly stations you will be escorted to your lifeboats/liferafts.

.5 All passengers are requested to carefully study the safety instructions behind their cabin doors.

.6 All passengers are requested to follow the escape routes, shown.

.7 Do not use lifts/elevators.

.8 All passengers are requested to strictly obey the instructions given by the officers or crew.

.9 When you hear the abandon ship alarm which consists of one prolonged and one short blast repeated continuously, please act in the same manner as under the general emergency alarm.

.10 During the voyage you may hear some other sound signals. These are exclusively for the information of the crew. Please, act only if you hear the general emergency alarm or the abandon ship alarm.

.11 If you have any questions regarding safety, do not hesitate to ask any of the officers or crew.

B4/2.2 Briefing on how to dress and what to take to assembly stations

.1 Take your lifejacket and a blanket. You will find your lifejacket under your bed.

.2 Put on warm clothing, long-sleeved shirts, strong shoes and head covering whatever the weather. No high-heeled shoes.

.3 Do not forget personal documents, your spectacles and medicine if necessary.

.4 Do not return to your cabin to collect your property.

B4/2.3 Performing roll call

.1 At your assembly station one of the officers/crew will perform a roll call.

.4 각자의 집합장소로부터 구명정/구명벌까지는 인도될 것임.
.5 모든 여객은 선실의 출입문 뒤쪽에 있는 안전 지시 사항을 주의 깊게 숙지해야 함.
.6 모든 여객은 안내된 탈출 통로를 따라가야 함.
.7 승강기/엘리베이터는 사용하지 마시오.
.8 모든 승객은 사관/승무원이 내린 지시에 철저히 따라야 함.
.9 장성 1발과 단성 1발을 연속하여 반복하는 퇴선경보를 들으면, 일반 비상경보 하에서와 같은 요령으로 행동하시오.
.10 항해 중에 각자는 다른 음향 신호를 들을 수도 있음.
이는 전적으로 승무원에게만 알리기 위한 것임.
각자는 일반 비상경보나 퇴선경보를 들었을 때에만 행동을 취하시오.
.11 각자는 안전에 관하여 어떤 질문 사항이 있으면, 사관이나 승무원 누구에게나 주저하지 말고 묻기 바람.

B4/2.2 복장의 착용법 및 집합장소에로의 지참물에 대한 설명

.1 각자의 구명동의와 담요를 지참하시오.
각자는 각자의 침대 밑에서 구명동의를 발견할 것임.
.2 날씨에 관계없이 보온용 의복, 긴소매 셔츠, 단단한 신발과 머리 보호구를 착용하시오. 굽 높은 구두는 안 됨.
.3 개인의 서류, 각자의 안경과 필요한 경우 의약품을 잊지 마시오.
.4 각자의 소지품을 가지러 선실로 돌아가서는 안됨.

B4/2.3 인원 점검의 실시

.1 각자의 집합 장소에서 사관/승무원 중 한 명이 인원점검을 실시할 것임.

.2 The officer/crew will say "This is a roll call", and s/he will call out the passengers individually by their names.
.3 When your name is called out, please answer loudly "Here".
.4 If one of your cabinmates is not able to attend the roll call, please inform the officer/crew immediately.

B4/2.4 Briefing on how to put on lifejackets

.1 *(dependent on type of lifejacket used)*
- pull the lifejacket over your head
- tighten the strings well
- pull the strings around your waist and tie in front.

.2 Follow closely the demonstration given by the officer/crew. The crew members will help you if necessary.
.3 Carefully study the demonstration in the pictures in your cabins.
.4 Carefully study the demonstration in the diagram at the assembly station.

B4/2.5 Instructions on how to embark and behave in lifeboats/ liferafts

.1 Enter the lifeboat/liferaft only when ordered by an officer/ lifeboatman.
.2 Clear the entrance of the lifeboat/liferaft immediately after entering.
.3 Do not push each other when entering the lifeboat/ liferaft.
.4 Hold on to ropes or to your seat when lowering/hoisting.
.5 Sit down in the lifeboat/liferaft immediately.
.6 Keep your lifejackets on.
.7 Provisions and drinking water will be distributed by an officer/lifeboatman only.
.8 Strictly obey all instructions given by the officer/ lifeboatman.
.9 Discipline in the lifeboat/liferaft is of vital importance.

.2 사관/승무원은 “이는 인원 점검입니다.”라고 말한 뒤, 승객의 이름을 개별적으로 호명할 것임.
.3 각자의 이름이 호명되면 “Here"라고 큰소리로 대답하시오.
.4 같은 선실의 승객 중 한 명이 인원 점검에 참석할 수가 없을 경우, 즉시 사관/승무원에게 알리시오.

B4/2.4 구명동의의 착용법에 대한 설명

.1 (사용되는 구명동의의 형식에 따름)
- 각자의 머리위로 구명동의를 끌어올리시오.
- 끈을 충분히 단단하게 하시오.
- 끈을 각자의 허리 주위로 둘러서 앞쪽에 매시오.

.2 사관/승무원이 보인 시범에 잘 따르시오.
필요하면 승무원이 각자를 도울 것임.
.3 각자의 선실 내에 있는 그림의 시범 내용을 주의 깊게 익히시오.
.4 집합장소에 있는 도해의 시범내용을 주의 깊게 익히시오.

B4/2.5 구명정/구명벌에의 탑승 요령과 행동 요령에 관한 지시

.1 오로지 사관/구명정수에 의하여 지시될 때에만 구명정/구명벌에 들어가시오.
.2 구명정/구명벌에 들어간 후에는 즉시 입구를 깨끗이 하시오.
.3 구명정/구명벌에 들어갈 때 서로 밀지 마시오.
.4 내릴 때/올릴 때 로우프나 각자의 좌석을 붙잡으시오.
.5 구명정/구명벌 내에서는 즉시 착석하시오.
.6 각자의 구명동의를 착용하고 있으시오.
.7 식량과 음료수는 오로지 사관/구명정수에 의해서만 분배됨
.8 사관/구명정수로부터 하달되는 모든 지시에 철저히 순종하시오.
.9 구명정/구명벌 내에서는 규율을 지키는 것이 지극히 중요함.

B4/2.6 On-scene measures and actions in lifeboats/liferafts

.1 Keep a sharp lookout for persons in the water.
.2 Have a line/hook/knife/lifebuoy ready.
.3 Do not take off your shirts/long trousers/head covering whatever the weather.
.4 Pump out the water/free the lifeboat/liferaft from water.
.5 Who needs medical first aid?
.6 Everybody will get the same ration of provisions and water.
.7 Warning! Do not drink sea water whatever the situation.
.8 We will send a MAYDAY.
.9 We will fire rockets/use smoke buoys/… to attract attention.
.10 We will join the other lifeboats/liferafts.

B4/3 Attending to passengers in an emergency

B4/3.1 Informing on present situation

.1 The vessel was abandoned in position … due to fire/grounding/collision/flooding/heavy list/serious damage/ ….
.2 Keep calm. There is no reason to panic.
The officers/lifeboatmen know exactly what to do.
.3 There are enough life-saving appliances for everyone on board.
.4 The Maritime Rescue Co-ordination Centre/vessels in the vicinity have already been informed of our situation.
.5 Vessels/helicopters/aircraft are coming to our rescue.
.6 Vessels/helicopters/aircraft will reach us within … hours.
.7 We have radio contact with rescue craft.
.8 There are enough provisions and drinking water for 48 hours.
.9 You obtain medicine for seasickness from the lifeboatman.

B4/2.6 구명정/구명벌 내에서의 현장 조치 및 행동

.1　물 속의 사람에 대하여 엄중한 견시를 유지하시오.
.2　줄/갈고리/칼/구명부환을 준비하시오.
.3　날씨가 어떠하든 각자의 셔츠/긴 바지/머리 보호구를 벗지 마시오.
.4　구명정/구명벌로부터 물을 펌프로 퍼내시오/물을 제거하시오
.5　누가 의료 응급처치를 필요로 하는가?
.6　각자는 모두 같은 양의 식량과 음료수를 분배받을 것임.
.7　경고! 어떤 상황 하에서도 해수를 마시지 마시오.
.8　우리는 메이데이를 발신할 것임.
.9　우리는 주의를 끌기위해 로켓 발사/발연부표를 사용/…할 것임.
.10　우리는 다른 구명정/구명벌들과 합류할 것임.

B4/3 비상시의 승객 관리

B4/3.1 현재 상황의 통보

.1　화재/ 좌초/ 충돌/ 침수/ 큰 횡경사/심한 손상/…으로 인하여 위치 …에서 퇴선하였음.
.2　조용히 하시오. 당황할 이유는 없음.
사관/구명정수는 무엇을 해야할지를 정확히 알고 있음.
.3　선상의 모든 사람에게 충분한 구명설비가 있음.
.4　우리의 상황에 대하여 해상구조조정본부/근처의 선박들에게 이미 통지되었음.
.5　선박/헬리콥터/비행기들이 우리를 구조하기 위해 오고있음.
.6　선박/헬리콥터/비행기들이 …시간 이내에 도착할 것임.
.7　우리는 구조선과 무선 교신을 하고 있음.
.8　48시간 동안에 대한 충분한 식량과 음료수가 있음.
.9　구명정수로부터 배멀미 약을 받으시오.

B4/3.2 Escorting helpless passengers

.1 … persons are missing.
.2 Search all cabins/WC/showers for missing persons.
.3 Assist those who need help.
.4 Help children, elderly, disabled, injured or sick persons with lifejackets.
.5 Give assistance when entering lifeboats/liferafts.
.6 We require warm clothing and blankets for the children/elderly/disabled/injured/sick.
.7 We require a stretcher for the disabled/injured/sick.
.8 All persons, please move closer.
.8.1 The elderly/disabled/injured/sick need room to lie down.
.9 Everyone, please, be quiet. The children/the sick need rest.

B4/3.2 거동이 불편한 승객의 보호

.1 …명의 사람이 없어졌음.
.2 모든 선실/ 화장실/ 샤워실에 대하여 없어진 사람을 수색하시오.
.3 도움을 필요로 하는 사람을 원조하시오.
.4 어린이, 연장자, 불구자, 부상자, 병자가 구명동의를 착용하도록 도우시오.
.5 구명정/구명벌에 들어갈 때 도움을 주시오.
.6 어린이/ 연장자/ 불구자/ 부상자/ 병자용으로 보온복과 담요 가 필요함.
.7 불구자/부상자/병자용으로 들것이 필요함.
.8 모든 사람은, 더 다가가시오.
.8.1 연장자/불구자/부상자/병자가 누울 장소가 필요함.
.9 여러분, 정숙하시오. 어린이/병자가 휴식을 필요로 함.

주 해(註解)

1) SMCP를 사용하여 어떤 의사 표시를 하였더라도 그 의사 표시 된 내용이 국제해상충돌예방규칙, 각 지방의 특별법, 권고, IMO의 항로 지정 등에 정하여진 내용과 어긋날 경우에는 의사 표시의 내용보다 위의 법, 규정이 우선한다는 취지임.

2) IMO STCW 협약 표 A-Ⅱ/1, 총톤수 500톤 이상의 선박에서 항해 당직을 담당하는 해기사에 대한 최저 기준에 "타선 또는 해안국과 교신하고 다국적 언어를 사용하는 선원과 함께 해기사의 임무를 수행할 수 있게 하는 영어에 관한 적절한 지식과 IMO 표준 해사 통신 용어를 이해하는 능력"이 포함되어 있음. 또 SOLAS 제4장 제16규칙을 참조할 것.

3) STCW 협약의 첨부서2, 선원의 훈련, 자격 증명 및 당직 근무 코드는 Part A에 STCW 협약 부속서의 규정에 관한 강제 기준이, Part B에 STCW 협약과 그 부속서의 규정에 관한 권고 지침이 정해져 있음.

4) 서로 다른 언어를 사용하는 다국적 선원들이 승선하는 선박 내에서 사용될 수 있는 대화 내용이 수록되어 있음.

5) Code(식별어)의 밑줄 친 부분에 강세가 있음.

6) 국제신호서의 숫자 통화표(figure-spelling table)에 있는 식별어와 본 SMCP의 식별어가 다른 경우는 본 SMCP의 내용에 따른다.

7) ITU의 RR 또는 전파법(국내)을 참조할 것.

8) Universal Time Coordinate(UTC).
A time scale maintained by the International Time Bureau.

For most practical purpose UTC is equivalent to mean solar time at the Prime Meridian i.e. GMT.

9) may의 가정법
might + root : ~해도 좋으련만.
You might ask before you use my typewriter.

10) ITU의 RR, 국내의 전파법, 또는 본서의 부록을 참조할 것.

11) 실무에서는 보통 gangway라고 하는 경우가 많다.

12) 선박의 마스트나 연돌 등의 장애물에 의하여 레이더의 스캐너에서 나온 전파가 차단되어 목표물까지 진행하지 못하는 구간으로서 스캐너에서 나온 전파가 전부 차단되는 구간을 blind sector, 전파의 일부가 차단되는 구간을 shadow sector라 한다.

13) 소형의 bulldozer.

14) 부표의 형태나 색상 등에 관하여는 IALA Buoyage System, Uniform System 등에서 정하는 바를 참조할 것.

15) IMDG Code의 "Requirement concerning transport operation" 편에 화물의 종류별 적부 요건이 명시되어 있음.

16) 항해일지의 Noon Position Data란에 기재하는 직항침로(course made good) 와는 개념상 차이가 있음을 주지할 것.

17) 원유 양하시에 양하되고 있는 원유의 일부를 본선의 원유 탱크에 고압으로 분사하여 탱크의 벽 등에 부착된 고형의 침전물을 용해시켜 양하함으로서 탱크를 세정하는 방법.

18) 기본수준면(Datum level for sounding)으로서 약최저저조면(Approximate lowest low water)을 말함.

19) 전세계 해상 조난 및 안전 시스템(GMDSS) 식별부호란 선박 통신장비에 의하여 발신될, 그리고 그 선박을 식별하기 위하여 사용될 해상이동업무 식별부호(MMSI), 선박호출부호, INMARSAT 식별부호 그리고 일련번호 식별부호(Serial number)를 말한다.

20) 항내 등에서 선박을 조종하는 경우에 닻을 투묘하고 해저를 따라 끌므로서 조종의 효율을 높이는 방법으로서, 대개 수심의 1.5~2배 정도의 닻줄을 신출하며, 위험물의 회피, 제자리 회두 또는 대각도 변침, 회두의 억제, 타력 억제 등의 목적으로 쓰인다.

21) 디지털선택호출(DSC)이란 무선국이 타국 또는 타국의 그룹과 연락을 취하고 정보를 송신할 수 있도록 하는 디지털 코우드를 사용하는 기술로 국제 무선 자문위원회(CCIR)의 관련 권고에 적합한 것을 말한다.
DSC는 GMDSS의 중추적인 역할을 하는 무선설비의 하나로서 HF, MF, VHF의 주파수를 사용하여 선박 및 해안국의 호출, 선박으로부터 조난통보의 발신, 해안국이 조난통보를 확실히 수신하였다는 것을 조난선에 알리는 수신 확인 통보, 선박 또는 해안국으로부터의 조난통보를 다른 선박이나 해안국에 중계하는 용도로 사용된다. (부록 참조)

22) 비상 위치 지시용 무선표지는 조난 발생시 조난자가 자기의 위치를 구조자에게 알리는 장치로서 지상계와 위성계 장치가 있다.
지상계는 2,182KHz, 121.5MHz, 243MHz, 156.525MHz 등을 사용하여 RDF등 지상계 무선 수신기로 조난자의 위치를 탐지하며, 위성계는 COSPAS-SARSAT system을 사용하는 것과 INMARSAT 위성을 사용하는 장치로 구분된다,

23) SAR 조약(해상에 있어서의 수색 및 구조에 관한 국제 조약)에 기초한 수색구조체제를 구축하기 위해서는 그에 적합한 효율적인 통보체제가 필요한데, 현대의 선박은 점차 자동화가 진행되어 소수의 인원으로 운항하게 됨에 따라 SAR시스템에서 요구되는 선상 탑재 시스템도 더욱 자동화된 장치로 변하게 되었다. 종래의 선박 조난 통보는 모르스 무선 전신이 주체가 되어 있었으나 SAR 시스템 하에서의 이러한 구난 통신 시스템은 인공 위 성이 포함된 현대적인 통신기술을 사용하는 신뢰성이 높고 자동화한 통신체제로 변하게 되었다. 이렇게 새로워진 GMDSS체제는 1992년 2월1일부터 적용되기 시작하여 1999년 2월 1일부터 전면적으로 실시되고 있다.

24) GPS의 위치 보정 시스템의 하나로서 우리나라 연안에는 8개의 DGPS국이 설치되어 운용중이고, 대부분의 선진국 해안에도 다수 설치되어 운용되고 있다. 위치의 정확도는 유효거리 100해리 이내에서 수 m 이하의 측위 오차를 보인다,

25) 자선의 항로로부터 이탈할 수 없도록 하는 작업에 종사하는 선박 즉, 예항에 종사하는 선박, 준설선, 수중작업선, 잠수작업선, 소해작업선 등을 말한다.

26) CONTENTS OF IMDG CODE

Chapter 2.1-Class 1-Explosives
- 2.1.1 Definitions and general provisions
- 2.1.2 Compatibility groups and classification codes
- 2.1.3 Classification procedure

Chapter 2.2 -Class 2- Gases
- 2.2.1 Definitions and general provisions
- 2.2.2 Class subdivisions
- 2.2.3 Mixtures of gases

Chapter 2.3 -Class 3- Flammable liquids
- 2.3.1 Definitions and general provisions

2.3.2 Assignment of packaging group
2.3.3 Determination of flash-point

Chapter 2.4 -Class 4- Flammable solids; substances liable to spontaneous combustion, substances which, on contact with water, emit flammable gases

2.4.1 Definitions and general provisions
2.4.2 - Class 4.1- flammable solids, self-reactive and related substances and desensitized explosives
2.4.3 - Class 4.2- Substances liable to spontaneous combustion
2.4.4 - Class 4.3- Substances which, on contact with water, emit flammable gases

Chapter 2.5 -Class 5- Oxidizing substances and organic peroxides

2.5.1 Definitions and general provisions
2.5.2 -Class 5.1- Oxidizing substances
2.5.3 -Class 5.2- Organic peroxides

Chapter 2.6 -Class 6- Toxic and infectious substances

2.6.1 Definition
2.6.2 -Class 6.1- Toxic substances
2.6.3 -Class 6.2- Infectious substances

Chapter 2.7 -Class 7- Radioactive materials

2.7.1 Definition

Chapter 2.8 -Class 8- Corrosives

2.8.1 Definition and properties
2.8.2 Assignment of packaging groups to corrosive substances

Chapter 2.9 -Class 9- Miscellaneous dangerous goods

2.9.1 Definition
2.9.2 Assignment of packaging groups to miscellaneous dangerous substance

27) 가연성 가스를 발생시키는 원유 등의 화물을 하역할 경우, 화물 탱크의 빈 용적에 가연성 가스와 공기가 혼합되어 폭발성의 기체를 형성할 위험이 있다. 이러한 위험한 상태가 되는 것을 피하기 위하여 선박 보일러의 배기가스를 몇 단계에 걸쳐 처리한 다음, 산소 농도가 적은 상태로 만들어 화물 탱크의 빈 용적에 주입함으로써 탱크 내의 대기가 안전한 상태에 있도록 하는 것을 말한다.

28) 선박이 항해할 때 바람이나 해조류에 밀려서 항적의 방향이 선수미선의 방향과 일치하지 않는 경우가 있는데, 이때 선박의 항적과 선수미선과의 교각을 풍압차라 하며, 선수미선을 기준으로 하여 +1도, -1도 등 각도 단위로 표시한다.

29) 전파법상 무선국의 업무 분류 중 하나로서, 해상이동업무라 하면 선박국과 해안국간, 선박국 상호간 또는 관계 선상 통신국 상호간의 무선통신업무를 말한다. 이에 참가하는 업무국으로서는 선박국, 해안국, 일반 해안국, 어업용 해안국, 선상 통신국 등이 있는데 이들 각 국의 식별을 위하여 부여되는 식별 부호로서 9자리의 숫자이다.

30) 우리나라의 해상 SAR 조직은 수난구호업무의 상설 총괄 조정기관인 중앙구조조정본부(Central Maritime Rescue Co-ordination Center: 해양경찰청), 5개의 상설 집행 기관인 구조조정본부(Maritime Rescue Co-ordination Center: 인천, 부산, 목포, 동해, 제주 해양경찰서), 7개의 구조지부(Rescue Sub-Center: 태안, 군산, 여수, 통영, 울산, 포항, 속초 해양경찰서) 및 구조대(Maritime Rescue Unit)로 구성되고, 중앙구조조정본부와 구조조정본부에는 협력기관 및 협력단체간의 유기적인 협조체제 구축을 위하여 중앙 및 지역 수난 구호 대책위원회가 있다. 그리고 미국 및 캐나다의 경우는 각 지역의 coast guard, 일본은 해상보안청의 각 관구 해상보안부가 이 역할을 한다.

31) 조종이 자유롭지 못한 상태(not under command)라 함은 본래

자력으로 조종이 가능한 선박이 예외적인 사정으로 말미암아 국제해상충돌예방규칙의 규정에 맞추어 뜻대로 조종 할 수 없는 선박 자체의 상태를 말한다.

32) 선박이 자선의 안전한 속력을 결정함에 있어서 고려해야 할 요인으로서는 당시의 시정, 어선이나 기타의 선박을 포함하는 교통의 밀도, 자선의 정지거리와 선회성능에 관계되는 조종성능, 육지의 등화 또는 자선의 등화의 반사광 등에서 생기는 배경광의 존재, 바람이나 해면 및 해류의 상태와 항행 상 위험의 근접, 수심과 흘수와의 관계, 레이더를 포함하는 항해기기의 성능 및 상태 등이다.

33) (a) expanding square search(for use by one ship)
(b) sector search(for use by one ship)
(c) parallel track(for use by two ships)
(d) parallel track(for use by three ships)
(e) parallel track(for use by four ships)
(f) parallel track(for use by five or more ships)
(g) ship/aircraft co-ordinated search(for use by ship and aircraft) 등이 있다.

34) oil boom, oil skimmer, oil suction device, oil dispersant, craft for dispersant application 등을 말함.

35) 길이 180cm 이상으로서, 보통 발판보다 긴 발판을 말한다. spreader와 spreader 사이의 간격은 발판 8개 이내이어야 한다.

36) 어느 탱크로부터 유류의 양하가 거의 완료되어 잔존 유량이 적을 경우, 공기 흡입 등으로 인하여 main cargo pump로는 잔존 유류를 양하할 수가 없다. 이러한 경우에 왕복동 펌프나 ejector 등의 stripping pump로 잔유를 양하 하는 것을 말한다.

37) "Underway" is not the same as "making way". "Making way"means the vessel is actually moving through the water either under her own power or with the help of tugs. Thus a vessel steaming at 12 knots is "underway" and "making way". A vessel which is 10 miles from land and stopped, but not at anchor, is "underway" but not "making way".

38) 미국에서는 Married fall, 영국에서는 Union purchase system이라 하고 때로는 Yard and stay system이라고도 불리워진다. 이 의장법은 하역 속력이 빠른 것이 장점인데 사용 도중에 양쪽 winch의 조종이 조화되지 않으면 화물의 swing이 생겨 fall과 guy에 과대한 장력이 걸려서 절단의 위험이 있으므로 주의해야 한다.

39) SMNV에서의 way point와는 그 개념이 달라졌으므로 주의를 요한다.

40) 대개, 통과 교통에는 사용되지 않으며 지역적인 특별 규칙이 적용될 경우가 있다.

41) 기관을 즉시 정지, 전진, 후진시킬 수 있는 상태에서 운전하는 경우의 속력으로서 통상 sea speed 보다 저속이며, 선박에 따라서는 manoeuvring speed로 운항하는 경우와 sea speed로 운항하는 경우에 사용하는 연료를 달리하는 선박도 있다.

42) 통항분리방안에 있어서의 외측 한계는 일방 통항이 요구되는 통항로의 외부 경계가 되며 이 경계 밖에서는 모든 선박이 어떤 방향으로든 항행 할 수가 있다. 분리대와 통항로의 길이와 폭은 당해 해역에 대한 제반 사항을 고려하여 정하여지지만 실제상 폭은 1~2 해리, 길이는 6~12 해리가 대부분이다.

43) 통항 분리 방안은 통항이 폭주하거나 혼잡한 해역에서 반대 방향 혹

은 거의 반대 방향으로 진행하는 통항을 분리함으로써 전손을 초래하는 충돌사고의 70%이상을 점하는 선박의 마주치는 상태의 발생 빈도를 감소시키고 충돌의 위험성을 줄이고자 하는 제도적 장치이다.

44) SOLAS 제5장의 제8-2규칙에 의하여, 정부는 해상 교통량 또는 발생 가능한 위험의 정도가 선박통항관제서비스(VTS)의 설립을 요구한다고 판단되는 해역에 VTS의 설립을 위한 필요한 조치를 취하여야 하며, 우리나라는 2000년 기준으로 11개 주요 항구에 VTS업무가 시행되고 있다.

45) 조난통신을 통제하는 구조조정본부, 수색 및 구조작업을 조정하는 기구 또는 동 작업에 참여하는 해안국은 그 통신에 혼신을 야기하는 국에게 침묵을 명할 수 있다. 이 지시는 상황에 따라 모든 국 또는 1개국에 발송할 수 있으며 어느 경우에 있어서도 무선 전화에 있어서는 프랑스어 “silence, m'aider로 발음되는 ”SEELONCE MAYDAY"라는 용어를 사용해야 한다.(부록 참조)

46) 호흡구(breathing apparatus)에는 공기 펌프식 호흡구와 자장식 호흡구가 있는데, 공기 펌프식은 요즈음 거의 쓰이지 않으며 자장식 호흡구로는 압축 공기를 내장하고 있는 용기를 사용자가 휴대하도록 된 형태가 많이 쓰인다.

47) 포말 소화기는 A급, B급 화재의 진화에 효과적이고 CO_2 소화기는 A, B, C급 화재에 효과적이며 기타 분말 소화기, 할론가스 소화기 등이 있는데 할론가스 소화기는 오존층 파괴의 원인이 된다는 이유로 사용이 금지되었다.

48) 선체에 틈이 생겨 침수가 시작되면 cement box에 의하거나, 용접에 의하거나 또는 방수판이나 방수 mat 등에 의하여 누수를 억제하고 pump로 물을 퍼내는 작업을 해야 되겠지만, 화물선에서는 수압이나 적재 화물 등으로 인하여 방수 작업의 실시가 어려운 것

이 사실이다. 따라서 SOLAS 규정에서도 화물선에는 침수 제어용 비품의 비치 의무를 부과하지 않고 있으며 침수 제어 훈련도 강제적으로 실시하도록 되어 있지는 않다.

49) 선박이 침몰하는 것을 방지하기 위하여 일부러 해안에 좌초시키는 것을 좌안(beaching)이라 한다. 좌안에 적합한 해안은 암석이 없고, 경사가 완만하며, 지반이 딱딱하여 해저면의 suction 작용이 없고, 강한 조류나 swell이 없고, 외해에 노출되지 않은 해안이다. 그리고 간만의 차는 어느 정도 있는 편이 좋은데 이는 간조 시에 선체의 손상부를 수리할 수도 있기 때문이다. 그리고 좌안은 만조 시나 만조 후 낙조가 시작할 때 실시하는 것이 좋은데, 이는 상당한 속력으로 선체를 올릴 필요 없이 해안에 걸리기만 하면 ebb tide에 의하여 좌안이 되기 때문이다.

50) "Piracy" is defined in the 1982 United Nations Convention on the Law of the Sea (UNCLOS) (article 101) as follows:

"Piracy consists of any of the following acts;

(a) any illegal acts of violence or detention, or any act of depredation, committed for private ends by the crew or the passengers of a private ship or a private aircraft, and directed:

(i) on the high seas, against another ship or aircraft, or against persons or property on board such ship or aircraft;

(ii) against a ship, aircraft, persons or property in a place outside the jurisdiction of any State;

(b) any act of voluntary participation in the operation of a ship or of an aircraft with knowledge of facts making it a pirate ship or aircraft;

(c) any act inciting or of intentionally facilitating an act described in sub-paragraph (a) or (b)."

"Armed robbery against ship" is defined in the draft Code of Practice for the Investigation of the Crimes of Piracy and Armed Robbery Against Ships(IMO MSC/Circ.984) (article 2.2) as follows;
"Armed robbery against ships means any unlawful act of violence or detention or any act of depredation, or threat thereof, other than an act of "piracy", directed against a ship or against persons or property on board such ship, within a State's jurisdiction over such offences."

51) 사람이 물에 빠졌을 경우에는
㈎ 선교에서는 물에 빠진 사람이 프로펠러에 말려 들어가지 않게 사람이 빠진 쪽으로 전타함 (kick현상을 이용)
㈏ 구명부환 등을 물에 빠진 사람 가까이에 던진다.
㈐ 즉시 선교에 알려서 비상배치와 필요한 조종을 하게 한다.
㈑ 물에 빠진 사람으로부터 계속 시선을 떼지 않는다,
㈒ 근처에 선박 왕래가 있으면 "O"기를 올리거나, 기적신호(장음 3회)를 울리거나, 긴급통보를 하여 타선박에게 이를 알린다.

52) SART(수색 구조용 레이더 트랜스폰더)는 선박 또는 항공기에 장비된 9 GHz 대의 레이더 전파에 응답하여 같은 9 GHz 대의 전파를 발사하며, 그것을 선박이나 항공기의 레이더에 20개의 점으로 이루어진 신호로 표시함으로써 조난자의 위치를 알 수 있는 장치이다. GMDSS에 있어서 SART는 SOLAS 협약 제4장 제6규칙에 정하여진 바와 같이 생존정에 항구적으로 취부 될 수도 있으며 float free로 동작하는 형식일 수도 있다.

53) 보우퍼트 풍력계급(Beaufort wind scale)은 Sir Francis Beaufort 제독이 1806년에 고안한 풍력의 계급으로서 풍속에 따라 0~12까지의 13등급으로 풍력을 구분한다.(부록 참조)

54) met. area는 해상기상예보해역으로서 전 세계의 해역을 I~XVI로 16분할하여 GMDSS 나 JMC의 해상 기상예보시에 이용되며 NAVAREA 와 동일한 분할 내용을 가진다.

55) 열대성 저기압으로서 그 발생 장소에 따라서 명칭을 달리하여, 경도 180도 선의 서쪽의 북태평양과 남지나해에서 발생하는 것을 Typhoon이라 하고, 북태평양의 경도 180도 선의 동쪽 해역과 멕시코만 및 카리브해 등에서 발생하는 것을 Hurricane, 오스트레일리아의 서방 해상 및 티모르해 부근에서 발생하는 것을 Willy-Willy라 하며, 인도양 및 남태평양 서부에서 발생하는 것을 Cyclone이라 한다.

56) 밀리바(mb)나 헥토파스칼(hPascal)이나 단위의 명칭은 다르나 그 수치는 동일하다. 예를 들면 1013 mb는 1013 hPascal이다. 예전에는 기압의 단위로서 밀리바를 사용해 왔으나 현재는 미터법인 헥토파스칼을 쓴다.

57) 태풍의 중심을 통과하는 진행 축에 대하여 우측에 있는 반원을 위험반원, 좌측에 있는 반원을 가항반원이라 한다. 위험반원에 비하여 가항반원의 해역에서 항해하기가 수월하며, 태풍권 내로부터 탈출하는 방법으로서는 RRR법칙, LRL법칙 등이 적용된다.

58) Tunami : 해저 지진이나 해저 화산 폭발에 의하여 생성된 물결이 사방으로 퍼져 큰 파도가 되어 해안에 도달하는 것을 말한다. 쓰나미의 주기는 수십분에서 수시간, 파장은 수백 km 정도가 되며, 파고는 해양상에서는 수십 cm 이하이므로 이의 관측도 어려우나 만, 특히 V형의 만 내에 도달하면 수위가 급격히 높아져서 피해를 입힌다.

59) 시정은 대기의 혼탁도를 나타내는 것으로서 이의 관측 요령은;

(1) 목표물의 형태가 인식되는 최대거리를 측정한다.

(2) 정상적인 시력의 사람이 나안으로 측정한다.

(3) 방향에 따라서 시정이 다른 때에는 최단거리 방향의 시정을 취한다.
(4) 야간에는 주간에 준하여 행한다. 시정은 밝기에는 관계되지 않는다.
(5) 시정 계급표는 0~9까지 10 단계의 계급이 있다.(부록 참조)

60) Pack ice(유빙) : 얼음 면적과 물 면적의 비율에 따라서 very close pack ice, close pack ice, open pack ice, very open pack ice 등의 등급이 있다.

Iceberg(빙산) : 해상에 부유하고 있는 큰 빙괴로서 해면상의 높이는 5m 이상의 것을 말하며 빙하나 ice-shelf에서 분열된 것이다.

61) GPS 위성은 A, B, C, D, E, F의 6개 궤도면에 각각 4개씩 24개의 위성이 배치되어 있으며 위성 번호는 A1~F4까지로 정해져 있다.

62) LORAN-C의 국명은 각 LORAN국의 GRI(Group Repetition Interval) 를 μsec로 나타내어 마지막 0을 제거한 숫자를 국명으로 사용한다.
하나의 주국에는 W, X, Y, Z 등 3~4개의 종국이 있다.

63) 0 Racon : Racon is a radar transponder which emits a characteristic signal when triggered by a ship's radar. The signal may be emitted on the same frequency as that of the triggering radar, in which case it is superimposed on the ship's radar display automatically.

64) Dangerous wreck : Wreck always partially submerged or sunken wreck dangerous to surface navigation(less tan 15 fathoms over wreck).

65) 인천항이나 파나마 운하 등에 설치된 갑문을 말함.

66) SOLAS Chap. V, Regulation 17 참조.

67) 도선사가 요구하는 경우, 직경 28mm 이상의 것으로서 선박에 견고하게 고정된 두 개의 man rope를 제공해야 한다.

68) SOLAS Chap. V, Regulation 17.

69) 항만에 따라서는 정해 놓지 않는 경우도 있으므로 연구된 자료, 예를 들면 일본해난방지협회의 자료를 참조한다.(윤점동, 선박조종의 이론과 실무, 세종출판사, p. 138, 1993)

70) Landing area는 헬리콥터가 착륙할 수 있는 구역으로서 H자와 그 주위에 원이 그려진 마크로 표시하며, pick up area는 헬리콥터가 착륙하지는 않지만 그 구역의 상방에 떠서 사람이나 물건을 올리거나 내리는 장소로서, 중심에 원형의 점과 그 주위에 원이 그려진 마크로 표시된다.

71) 쇄빙선의 원조를 받고 있는 경우뿐만 아니라 운하 통과 시 등에 있어서도 선수 선미에 탐조등을 설치해야 한다.

72) Adopted on 27 Nov 1977 (Assembly 20th session, Agenda item 9)

73) 상대방에게 국제해상충돌예방규칙에서 어긋나는 내용의 요청을 해서는 안 된다.

74) 선박의 운항과 관련되는 국가 조직으로서는 Flag State(기국, 선적국), Port State(항만국), Coastal State(연안국) 등이 있다.

75) 실제상으로는, rescue가 현재 존재하는 위험으로부터 인명이나 재화를 구출하는 의미로 사용되는 경향이 있는데 반하여, salvage는 위험 상황이 종료된 이후에 재화의 가치를 회복하기 위하여 선체나 화물을 인양, 예인하는 의미로 쓰이는 경우가 많다(1989년 해난구

조에 관한 국제협약 참조)

76) Set : 유향 - 각도로 표시, Drift : 유정 - 노트로 표시.

77) Gale warning(강풍경보) - 예상 최대풍력이 8~9 일 때 발표함.
Storm warning(폭풍경보) - 예상 최대풍력이 10 이상일 때 발표함.

78) Traffic separation scheme 관련 항법에 대한 조항임.

79) 우리나라의 경우, 도선법 제20조를 참조할 것.

80) 전파법(국내)의 조난통신, 긴급통신, 안전통신 편을 참조할 것. (부록 참조)

81) 조난자가 조난통보를 발신할 경우, 조난통보의 구성은 경보신호, 조난호출, 조난통보의 순서로 구성하여 발신해야 한다. 그러나 GMDSS 체제 하에서는 최초로 조난 통보를 발신하는 것은 DSC distress alert 등 자동 스위치에 의하는 경우가 많다. 최초 조난신호의 발신을 VHF Ch. 16(156.8 MHz) 등으로 하고자 할 경우에 대하여는 전파법의 관련 조항과 본서의 부록을 참조할 것.

82) 실제로는 선교 당직사관이 직접 bridge telegraph를 조작하는 경우가 대부분이다.

83) 선박이 전진하고 있는 상태에서 선박의 선미로부터 프로펠러를 보았을 때, 프로펠러가 시계방향으로 회전하면 right-hand propeller 이고, 반시계 방향으로 회전하면 left-hand propeller 이다.

84) 1 마력(horse power)은 0.7461 kW임.

85) 선박이 전진하고 있는 상태에서, 선미로부터 프로펠러를 보았을 때

본선의 우현 측의 프로펠러가 시계방향으로, 좌현 측의 프로펠러가 반시계 방향으로 회전하면 outward, 이와 반대이면 inward라 한다.

86) Ship's particulars, Ship's condition, Engine condition, Steering condition, Thruster particular, Turning circle data, Stopping characteristics 등의 자료를 기재하여 도선사 승선 시에 제시하는 카드를 말한다.

87) 선박에 따라서 차이가 있겠으나 full sea speed에서 manoeuvring speed로 변경하고자 할 때에는 통상 30분 이전에 이러한 취지를 기관실에 통지하는 것이 좋다.

88) 김우숙, 레이더항법, 해문출판사, pp. 106~112, 1985 참조.

89) **Up and down** : anchor cable 이 수면과 수직이 되고 닻이 해저를 떠나려는 상태.
Anchor aweigh : anchor 의 crown이 해저에서 떨어진 상태.

90) 닻이 해저에 완전히 박혀 발묘될 위험이 없으므로 투묘 작업을 완료해도 좋게 된 상태.(When the ship has settled down to her anchorage, is not dragging the anchor and has a steady strain on the cable, she is said to be "brought up".)

91) 선박의 정박 상태를 표시하는 주간 형상물로서 흑색이며 직경 0.6m 이상으로서 구형이다.

92) 파나마 운하 통과를 위하여 요구되는 설비로서 상, 하, 좌, 우에 모두 roller가 설치되어 현측으로부터 계류삭이 어느 방향으로 나가더라도 원활하게 출입할 수 있도록 되어 있으며 폐색 도삭기라고도 한다. (A heavy steel fitting equipped with rollers, through which mooring line pass.)

93) **Bitt** : Strong short post of wood or iron, usually fastened to a deck in pairs and named according to their use, such as towing-bitts, mooring-bitts etc.
Bollard : Large post of wood, stone or metal usually on quay, wharf or dockside, for making ships fast.

※ 갑판 상에 설치된 것으로서 두 개가 나란히 한 조가 되어 있는 것을 bollard, 단 한 개의 기둥만 서 있는 것을 bitt라 하기도 한다.

94) A berth is a space in a harbour where a ship stays for a period of time e.g. SBM, Dolphin, T-jetty, Pier, Dock, Quay, Wharf, Anchorage. (윤점동, 선박조종의 이론과 실무, 세종출판사. pp.116~121, 1993 참조.)

95) NAVTEX는 원양 항해용의 NAVAREA와는 다르게, 연해 해역을 항행하는 모든 선박이 필요로 하는 통상의 기상예보, SAR 정보, 항행경보 등을 518 kHz로 방송한다. 수신기는 통보를 선택할 수가 있으며 우리나라에서는 변산반도와 죽변에 송신국이 있고 수신형태는 TELEX로 인쇄되어 출력되는 형태이다.

96) MAREP(Marine Report) : Position, course, speed and next port of call are some of the information vessels will give in a MAREP.
POSREP(Position Report): Usually passed by merchant vessels in Traffic Separation Schemes.

97) 복무수칙은 선교나 해도실에 비치하는데, 이는 선장이 선박운항의 책임을 맡은 당직사관에게 침로, 속력, 특기사항 등 필요한 내용을 숙지하도록 지시하는 것이다. 당직 항해사는 당직에 임하기 전에 복무수칙을 숙지하고 이에 서명해야 한다.

98) Track way type gravity davit에서 boat winch나 boat fall의 교체, 수리시에 대비트를 고정시키는 장치.

99) Slip hook, toggle pin, lock piece, pivot hook, release hook 등 구명정 이탈 장치를 말함.

100) 팽창식 구명벌의 팽창시에 trigger 역할을 하는 로프로서 구명벌과 본선의 선체와 연결되어 있어야 한다.

101) 소방원 장구(firemen's outfit)

(1) 화상을 예방할 수 있는 방화복.

(2) 고무 또는 절연성 재료로 된 방화 장화 및 방화 장갑.

(3) 충격으로부터 보호할 수 있는 헬멧.

(4) 최소 3시간 점등 될 수 있는 휴대용 전등.

(5) 주관청이 인정하는 방화용 도끼.

(6) 승인된 형식의 호흡구.

(7) 충분한 강도 및 길이와 내화성을 갖는 구명줄.

102) 국제 신호서에 따른 JF, JG, JH 신호, 또는 국제해상 충돌예방규칙 제 30조 참조.

103) International Aeronautical and Maritime Search and Rescue Manual.

104) Rescue boat(구조정)과 Life boat(구명정)의 일반요건에 대하여는 참고 서적(문채식, 해상안전론, 해문출판사, pp. 75, 121, 1996) 참조.

105) **Grain capacity** : 광석, 곡물, 석탄 등을 선창 내에 bulk로 적재할 때 기준이 되는 선창 용적이다.

Bale capacity : bale등 포장된 화물을 선창 내에 실을 때 기준이 되는 선창 용적을 말한다.

106) Twist lock, vertical stacker, bridge connector, turn buckle, corner hook, container lock, lashing rod 등을 말한다.

107) 쌀, 어분, 석탄, 면화, 금속분 등은 산화에 의하여 스스로 발열하여 자연 발화할 위험이 있다.

108) 유류 등이 갑판에 흘러 떨어질 경우 이를 받아 담아둠으로서 유류가 선외로 누출되는 것을 방지하도록 설치된 용기로서, 주로 탱커의 manifold 아래나 bunker 수급용 파이프 연결부의 아래에 설치되어 있다.

109) 선체의 전위와 부두의 전위가 서로 다른 상태에서 유조선의 manifold와 육상의 pipe line을 연결하게 되면 연결 순간에 방전이 일어나 위험한 상황이 전개될 수가 있다. 이러한 상황을 피하기 위하여, 유조선에서 선박의 manifold와 육상의 pipe line 연결하기 이전에 미리 도전성의 철사로 양자를 연결하여 전위차를 없애는데 이런 용도로 쓰이는 선을 bonding wire라 한다,

110) 액체 화물선의 manifold와 육상의 pipe line을 연결할 때 사용하는 육상의 연결용 pipe 임. (Chicksan arm이라고도 한다.)

111) Slop tank : 유조선의 탱크를 세정할 경우에 세정된 탱크로부터 나오는 유류와 세정수의 혼합물을 저장하는 탱크.
Sludge : 원유 등 유류의 침전물.

112) The Clip-on Unit is a portable version that is designed to fumigate cargo loaded in a shipping container. The unit can be fitted to the front of a standard 20 or 40 ft shipping container. This allows cargo to be fumigated

without the need for it to be unloaded.

113) 선원은 선주의 대리인으로서 성실하고 적절하게 화물을 관리해야 하며 그러한 내용을 항해일지에 상세히 기재함으로써 due regard 를 다했다는 증거 자료로 삼아야 한다.

114) SOLAS Chap.Ⅲ, Regulation 19 참조.

115) SOLAS Chap. Ⅲ, Regulation 6, 5.2항 참조.

부 록 (附錄)

부록 1

[1] VHF 무선전화 통신 절차

1. VHF 무선전화

VHF 무선전화는 대부분의 선박에서 가장 자주 사용하는 기본적인 통신설비이나, 통신 가능거리가 약 30마일로 비교적 짧다는 한계를 가지고 있다.

VHF 무선전화는 조난통신 채널이자 호출응답 채널인 Ch.16과, DSC에 의한 조난통신 및 호출응답 채널인 Ch.70을 포함하고 있다.

선박이 항해중일 때에는 항상 채널 16을 청수하고 있어야 하며, 항계 내에 들어올 때에는 지정된 항무통신 채널을 수신하여 한다. 그러나 VHF 무선전화가 2대 설치되어 있는 경우에, 한 대는 채널 16에, 다른 한 대는 항무통신 채널에 고정시켜 두는 것이 바람직한 청수 방법이다.

VHF 채널은 단신채널 (Simplex channel)과 복신채널 (Duplex channel)이 있는데, 단신 채널은 하나의 주파수로 송수신을 하기 때문에 송신과 수신을 번갈아 가며 해야 하는 불편함이 있다. 그러나 복신채널인 경우에는 송신 주파수와 수신주파수가 분리되어 있어 송신과 동시에 수신을 할 수가 있다. 그러나 복신 채널은 Ship-to-Ship 통신에서는 사용할 수 없다.

2. VHF 무선전화 통신방법

호출·응답 주파수인 채널 16으로 호출하는 것을 원칙으로 한다.

1) 일반적인 호출방법

☞ 호출은 다음 순서로 시행한다.

-. 피 호출국의 명칭/호출부호 또는 다른 식별표시	3회 이하
-. THIS IS	1회
-. 호출국의 명칭/호출부호 또는 다른 식별표시	3회 이하

☞ 그러나 연락 설정 조건이 양호할 때에는 다음과 같이 할 수 있다.

-. 피 호출국의 명칭/호출부호 또는 다른 식별표시	1회
-. THIS IS	1회
-. 호출국의 명칭/호출부호 또는 다른 식별표시	2회

☞ 연락이 설정된 후에는 명칭/호출부호 또는 다른 식별표시는 1회로 할 수 있다.

☞ 호출 및 통신 준비신호는 조난이나 긴급 및 안전의 경우를 제외하고는 채널 16에서 1분을 초과해서는 안 된다.

Establishing Contact - First Call
If your first call is not answered, proceed as follows: • second call after 2 minutes • third call after another 2 minutes • then wait for 3 minutes and repeat the procedure.
Name of station called (max. 3x; when conditions are good: max. 1x) + call sign This is Name of calling station (max. 3x; when conditions are good: max: 2x) + call sign Reason for calling Change to working channel … OVER
After first contact with the station called
Name/call sign of the station called This is Name/call sign of the station calling Text of message OVER or OUT

2) 호출의 반복

☞ 피 호출국이 응답하지 아니할 때에는 3분 간격으로 반복할 수 있다

☞ 호출을 재개하기 전에 피 호출국이 다른 국과 통신 중이 아닌 것을 확인하고, 그 호출이 진행 중인 통신에 혼신을 주지 않음을 미리 확인해야 한다.

3) 응답 형식

선박국이 채널 16으로 호출을 받은 경우, 호출국이 다른 주파수를 지정하지 않

는 이상 그 채널로 응답한다.

☞ 채널 16에서 연락 설정 조건이 양호할 때에는 다음과 같이 할 수 있다.

-. 피 호출국의 명칭/호출부호 또는 다른 식별표시 1회
-. THIS IS 1회
-. 호출국의 명칭/호출부호 또는 다른 식별표시 2회

☞ 연락이 설정된 후에는 명칭/호출부호 또는 다른 식별표시는 1회로 할 수 있다.

3. General VHF Radio Conduct

You should be aware that *all bridge conversations could be used as evidence* for inappropriate ship operation including the *Maritime English performance* involved in case of an incident, accident, distress, etc.

Voyage Data Recording
VHF communications relating to the ship operations **should be recorded.** One or more microphones positioned on the bridge should be placed so that **conversations at or near the conning station, radar displays, chart tables etc.,** are adequately recorded. As far as practicable the positioning of microphones should also capture intercom, public address systems and audible alarms on the bridge The time for which all stored data items are **retained should be at least 12 hours.** Data items which are older than this may be **overwritten with new data.**

1) Order of Priority in VHF Communications

Order of priority	
Distress communication preceded by	**MAYDAY**
Urgency communication preceded by	**PAN-PAN**
Safety communication preceded by	**SECURITE**
Any other communications, e.g. Routine communication (VTS communication, Pilot request, Tug assistance 등) General communication.	
Note: Stations **must stop using any VHF channel** when this is needed **for communications of a higher priority.**	

2) VHF Radio 사용원칙

General principles	
1	Do not make **unnecessary** transmissions.
2	Always follow the prescribed **VHF Radio Regulations and procedures.**
3	Speak **more slowly than in normal conversation,** especially when the addressee is writing down your message.
4	Maintain a **constant voice level.**
5	Pronounce **each word clearly** and spell it if necessary.
6	Always follow established Maritime English practice, particularly the application of **Message Markers and IMO SMCP.**
7	Always be **concise and unambiguous.**
8	**Use fixed formats,** e.g. MAREP etc., where applicable. *The application of MAREP will especially be dealt with.*

Before transmitting	
1	Make sure that **your call is really necessary.** *Do not jam VHF channels with trivial communications.*
2	**Plan what you want to say** before starting, write it down if necessary.
3	**Check** that the VHF set is switched to the **correct VHF channel.** *Check **working channels of coast stations** in the latest ITU List of Coast Stations.*
4	If you want to call **another ship, tell that ship which working channel** should be used for the exchange. *It will save problems later if, before you choose a working channel, you make certain that it is not already in use.* *First listen - then speak.*
5	**Do not interrupt other stations VHF exchanges.** *It can be dangerous to interrupt VHF communication in progress; it is even **illegal to interrupt** Distress and Urgency traffic.*
6	Do not make **unauthorized transmissions.** *It is forbidden to:* *·transmit false or misleading signals,* ***·transmit without identification*** *·make unnecessary or superfluous transmissions.*

4. 조난· 긴급· 안전통신

조난·긴급·안전통신을 함에 있어서는 특히 Radio Regulations의 준수가 가장 중요하다. 이러한 규정을 위반하는 것은 때때로 귀중한 시간을 낭비하거나, 재산이나 해양환경의 손실, 심지어는 인명의 손실을 초래할 수 있다.

또한 가능한 통신문을 짧게 하기 위하여 선박의 종류, Motor Vessel(MV), Motor Tanker(MT), Ferry 등과 같은 내용은 생략한다.

1) Scheme for a Distress Call / Message

Transmission of a Distress Call and a Distress Message International Distress and Calling Frequencies: VHF Channel 16 = 156.8 MHz and 2182 kHz
MAYDAY indicates *that a ship, an aircraft, or other vehicle or a person is threatened by grave and imminent danger and requires immediate assistance.* *Note:* ***In the DSC*** *a Distress Alert made on the appropriate* ***frequency/ VHF Ch 70*** *is automatically sent.*
Distress Call
MAYDAY **(3x)** this is **name of the vessel in distress** (3x) **call sign** *(MMSI if the alert was sent by DSC)*
Distress Message (순서)
MAYDAY **name of the vessel in distress** **call sign or other identification** (MMSI if the alert was sent by DSC) **+ position** **+ nature of distress** **+ assistance required** **+ any other useful information, such as:** - Master's intention - type of cargo (if dangerous) - weather and sea condition - time of abandonment - number and type of survival craft launched - number of persons abandoning/staying on board - details of location aids in survival craft or sea. **over**

2) Scheme for acknowledging a Distress Message

Acknowledgement of a Distress Message
*A vessel, **upon receiving a distress call/message** from another vessel in vicinity **should acknowledge receipt immediately** unless in sea areas A1/A2 (see below), when time (ca 5 minutes) should be given for a coast station/MRCC to acknowledge. Where the station in distress is a **long distance away**, then **acknowledge and relay to the nearest coast station**/MRCC only If no other acknowledgement has been heard.*
MAYDAY name/call sign of the **vessel in distress** this is name/call sign of the **acknowledging station** ***received MAYDAY***
Having acknowledged the distress call then continue as follows:
MAYDAY name of the **vessel in distress** (3x) call sign this is name of **assisting vessel** (3x) call sign **+ position of assisting vessel** **+ speed to distress position** **+ ETA at distress position** *(if advisable: bearing of the vessel in distress)* **over**
MAYDAY MAYDAY MAYDAY this is *Ringo Ringo Ringo call sign PPAY*
MAYDAY Ringo / PPAY *Position 62° 11.8' North, 007° 44' East.* *I am on fire after explosion in engine room.* *I am not under command.* *Fire is not under control.* *I require fire fighting assistance.* *My cargo is wood pulp.* *I have launched one lifeboat with 8 persons.* *3 officers and master staying on board.* *Warning: Many sailing yachts near distress position.* ***over***

부록 2

IMO Resolution A.954 (23).
(Adopted on 5th December 2003)
Proper use of VHF Channels at Sea

(본문 내용) 해상에서의 VHF 채널의 적절한 사용

해상에서의 VHF를 통한 무선통신은 항해의 안전에 있어 매우 중요하며, 잘못된 사용으로 인해 큰 사고를 야기할 수도 있음.

각 회원국 및 해상에서 VHF를 사용하는 선원들에게 VHF 무선통신의 적절한 사용에 관한 지침을 제공하여 해상에서의 인명의 안전을 증진시키고, 각 회원국에서 이에 대한 주의 및 적절한 조치를 취하기를 요청함.

동 결의문은 기존의 결의문 A.474(XII)를 대체함.

(Annex) GUIDELINES ON THE USE OF VHF AT SEA

(요약) VHF 통신의 기술 (청취, 통신의 원칙, 반복, 송신출력의 감소, AIS의 사용, 육상국과의 통신, 다른 선박들과의 통신, 조난통신, 호출, 채널의 변경, 스펠링, 호칭, 당직), VHF 통신절차(호출, 메시지의 교환), 표준메시지(IMO 표준해사통신용어, SMCP의 사용) 등에 대하여 규정하고 있음.

1. VHF COMMUNICATION TECHNIQUE

1.1 Preparation

Before transmitting, think about the subjects which have to be communicated and, if necessary, prepare written notes to avoid unnecessary interruptions and ensure that no valuable time is wasted on a busy channel.

1.2 Listening

Listen before commencing to transmit to make certain that the channel is not already in use. This will avoid unnecessary and irritating interference.

1.3 Discipline

VHF equipment should be used correctly and in accordance with the Radio Regulations. The following in particular should be avoided:

(a) calling on channel 16 for purposes other than distress, and very brief safety communications, when another calling channel is available;

(b) non - essential transmissions, e.g. needless and superfluous signals and correspondence;

(c) communications not related to safety and navigation on port operation channels; communication on channel 70 other than for Digital Selective Calling;

(d) occupation of one particular channel under poor conditions;

(e) transmitting without correct identification;

(f) use of offensive language.

1.4 Repetition

Repetition of words and phrases should be avoided unless specifically requested by the receiving station.

1.5 Power reduction

When possible, the lowest transmitter power necessary for satisfactory communication should be used.

1.6 Automatic identification system (AIS)

AIS is used for the exchange of data in ship-to-ship communications and also in communication with shore facilities. The purpose of AIS is to help identify vessels, assist in target tracking, simplify information exchange and provide additional information to assist situational awareness. AIS may be used together with VHF voice communications.

AIS should be operated in accordance with Resolution A.917 (22) as amended by Resolution A.956 (23) on Guidelines for the onboard operation use of shipborne automatic identification systems.

1.7 Communications with coast stations

On VHF channels allocated to port operations service, the only messages permitted are restricted to those relating to the operational handling, the movement and safety of ships and, in emergency, to the safety of persons, as the use of these channels for ship-to-ship communications may cause serious interference to communications related to the movement and safety of shipping in port areas.

Instructions given on communication matters by shore stations should be obeyed.

Communications should be carried out on the channel indicated by the shore station. When a change of channel is requested, this should be acknowledged by the ship.

On receiving instructions from a shore station to stop transmitting, no further communications should be made until otherwise notified (the shore station may be receiving distress or safety messages and any other transmissions could cause interference).

1.8 Communications with other ships

VHF Channel 13 is designated by the Radio Regulations for bridge to bridge communications. The ship called may indicate another working channel on which further transmissions should take place. The calling ship should acknowledge acceptance before changing channels.

The listening procedure outlined above should be followed before communications are commenced on the chosen channel.

1.9 Distress communications

Distress calls/messages have absolute priority over all other communications. When heard, all other transmissions should cease and a listening watch should be kept.

Any distress call/message should be recorded in the ship's log and passed to the master.

On receipt of a distress message, if in the vicinity, immediately acknowledge receipt. If not in the vicinity,

allow a short interval of time to elapse before acknowledging receipt of the message in order to permit ships nearer to the distress to do so.

1.10 Calling

In accordance with the radio regulations Channel 16 may only be used for distress, urgency and very brief safety communications and for calling to establish other communications which should then be conducted on a suitable working channel.

Whenever possible, a working frequency should be used for calling. If a working frequency is not available, Channel 16 may be used, provided it is not occupied by a distress call/message.

In case of difficulty to establish contact with a ship or shore station, allow adequate time before repeating the call. Do not occupy the channel unnecessarily and try another channel.

1.11 Changing channels

If communications on a channel are unsatisfactory, indicate change of channel and await confirmation.

1.12 Spelling

If spelling becomes necessary use the spelling table contained in the International Code of Signals and the radio regulations and the IMO Standard Marine Communication Phrases (SMCP)

1.13 Addressing

The words "I" and "You" should be used prudently. Indicate to whom they refer.

Example of good practice :
"Seaship, this is Port Radar, Port Radar, do you have a pilot?" **"Port Radar, this is Seaship, I do have a pilot."**

1.14 Watchkeeping

Every ship, while at sea, is required to maintain watches. Continuous watch keeping is required on VHF DSC Channel 70 and also when practicable, a continuous listening watch on VHF Channel 16. In certain cases Governments may require ships to keep a watch on other channels.

2 VHF COMMUNICATION PROCEDURES

2.1 Calling

When calling a coast station or another ship, say the name of that coast station once (twice if considered necessary in heavy radio traffic conditions) followed by the phrase THIS IS and the ship's name twice, indicating the channel in use.

Example :
"Port City, this is Sea Star, Sea Star, on Channel 14."

2.2 Exchange of messages

2.2.1 When communicating with a ship whose name is unknown but whose position is known, that position may be used. In this case the call is addressed to all ships.

Example :
"All Ships, this is Pastoria, Pastoria, Ship approaching number four(4) buoy, I am passing Belinda Bank Light."

2.2.2 Where a message is received and only acknowledgement of receipt is needed, say "received". Where a message is received and acknowledgement of the correct message is required, say "received, understood", and repeat message if considered necessary.

Example :
Message: Your berth will be clear at 08.30 hours. Reply: Received, understood. Berth clear at 08.30 hours."

2.2.3 Where appropriate, the following message should be sent:

"Please use/ I will use the IMO Standard Marine Communication Phrases ".

When language difficulties exist which cannot be resolved by use of the IMO Standard Marine Communication Phrases, the International Code of Signals should be used.

In this case, the word "INTERCO" should precede the groups of the International Code of Signals.

" Please use/ I will use the International Code of Signals "

2.2.4 Where the message contains instructions or advice, the substance should be repeated in the reply.

Example :
"Message : Advise you pass astern of me. Reply : I will pass astern of you."

2.2.5 If a message is not properly received, ask for it to be repeated by saying "**Say again**".

2.2.6 If it is necessary to change to a different channel say "**Change to channel ….**" and wait for acknowledgement before carrying out the change.

2.2.7 During exchange of messages, a ship should invite a reply by saying "**over**".

2.2.8 The end of a communication is indicated by the word "**out**".

3 STANDARD MESSAGES

3.1 Since most ship-to-shore communications are exchanges of information, it is advisable to use standard messages which will reduce transmission time.

3.2 Commonly used standard messages are given in the IMO Standard Marine Communication Phrases (SMCP), which should be used whenever possible.

REFERENCE DOCUMENTS

- 1974 SOLAS Convention, as amended, chapter IV on Radio communications.
- Radio Regulations, Appendix 18, Table of Transmitting Frequencies in the VHF Maritime Mobile Band.
- Resolution A.917(22), as amended by resolution A.956(23) on Guidelines for the onboard operational use of shipborne automatic identification systems (AISs).
- Resolution A.918(22) on IMO Standard Marine Communication Phrases (SMCP).

부록 3

[3] OPERATIONAL PROCEDURE FOR GMDSS DSC COMMUNICATIONS ON MF AND VHF

Introduction

Procedures for DSC communications on MF and VHF are described in § 1 to 4 below. The procedures for DSC communications on HF are in general the same as for MF and VHF.

1. Distress

1.1 Transmission of DSC distress alert

A distress alert should be transmitted if, in the opinion of the Master, the ship or a person is in distress and requires immediate assistance.

A DSC distress alert should as far as possible include the ship's last known position and the time (in UTC) when it was valid. The position and the time may be included automatically by the ship's navigational equipment or may be inserted manually.

The DSC distress alert is transmitted as follows:

- tune the transmitter to the DSC distress channel (2,187.5 kHz on MF, channel 70 on VHF) (see Note 1);

Note 1- Some maritime MF radiotelephony transmitters shall

be tuned to a frequency 1,700 Hz lower than 2,187.5 kHz, i.e. 2,185.8 kHz, in order to transmit the DSC alert on 2,187.5 kHz.

- if time permits, key in or select on the DSC equipment keyboard
  - the nature of distress,
  - the ship's last known position (latitude and longitude),
  - the time (in UTC) the position was valid,
  - type of subsequent distress communication (telephony),

  in accordance with the DSC equipment manufacturer's instructions;
- transmit the DSC distress alert;
- prepare for the subsequent distress traffic by tuning the transmitter and the radiotelephony receiver to the distress traffic channel in the same band, i.e. 2,182 kHz on MF, channel 16 on VHF, while waiting for the DSC distress acknowledgement.

1.2 Acknowledgement of a DSC distress alert

Acknowledgement of a DSC distress alert by use of DSC is normally made by coast stations only.

Only if no other station seems to have received the DSC distress alert, and the transmission of the DSC distress alert continues, the ship should acknowledge the DSC distress alert by use of DSC to terminate the call. The ship should then, in addition, inform a coast station or a coast earth station by any practicable means.

Ships receiving a DSC distress alert from another ship should defer the acknowledgement of the distress alert for a short interval, if the ship is within an area covered by one or more coast stations, in order to give the coast station time to acknowledge the DSC distress alert first.

Ships receiving a DSC distress alert from another ship shall :

- prepare for receiving the subsequent distress communication by tuning the radiotelephony receiver to the distress traffic frequency in the same band in which the DSC distress alert was received, i.e. 2,182 kHz on MF, channel 16 on VHF;
- acknowledge the receipt of the distress alert by transmitting the following by radiotelephony on the distress traffic frequency in the same band in which the DSC distress alert was received, i.e 2,182 kHz on MF, channel 16 on VHF:
  - "MAYDAY",
  - the 9-digit identity of the ship in distress, repeated 3 times,
  - "this is",
  - the 9-digit identity or the call sign or other identification of own ship, repeated 3 times,
  - "RECEIVED MAYDAY".

1.3 Distress traffic

On receipt of a DSC distress acknowledgement the ship in distress should commence the distress traffic by radiotelephony on the distress traffic frequency (2,182 kHz on MF, channel 16 on VHF) as follows:

- "MAYDAY"
- "this is"

- the 9-digit identity and the call sign or other identification of the ship,
- the ship's position if not included in the DSC distress alert,
- the nature of distress and assistance wanted,
- any other information which might facilitate the rescue.

1.4 Transmission of a DSC distress relay alert

A ship knowing that another ship is in distress shall transmit a DSC distress relay alert if

- the ship in distress is not itself able to transmit the distress alert
- the Master of the ship considers that further help is necessary.

The DSC distress relay alert is transmitted as follows:

- tune the transmitter to the DSC distress channel (2,187.5 kHz on MF, channel 70 on VHF),
- select the distress relay call format on the DSC equipment,
- key in or select on the DSC equipment keyboard:
    - All Ships Call or the 9-digit identity of the appropriate coast station,
    - the 9-digit identity of the ship in distress, if known,
    - the nature of distress,
    - the latest position of the ship in distress, if known,
    - the time (in UTC) the position was valid (if known),
    - type of subsequent distress communication (telephony);
- transmit the DSC distress relay call.

1.5 Acknowledgement of a DSC distress relay alert received from a coast station

Coast stations will, after having received and acknowledged a DSC distress alert, normally retransmit the information received as a DSC distress relay call, addressed to all ships, all ships in a specific geographical area, a group of ships or a specific ship.

Ships receiving a distress relay call transmitted by a coast station should acknowledge the receipt of the call by radiotelephony on the distress traffic channel in the same band in which the relay call was received, i.e. 2,182 kHz on MF, channel 16 on VHF.

The acknowledgement is transmitted as follows:

- "MAYDAY",
- the 9-digit identity or the call sign or other identification of the calling coast station,
- "this is",
- the 9-digit identity or call sign or other identification of own ship,
- "RECEIVED MAYDAY".

1.6 Acknowledgement of a DSC distress relay alert received from another ship

Ships receiving a distress relay alert from another ship shall follow the same procedure as for acknowledgement of a distress alert, i.e. the procedure given in § 1.2 above.

2. Urgency

2.1 Transmission of urgency messages

Transmission of urgency messages shall be carried out in two steps:

- announcement of the urgency message,
- transmission of the urgency message.

The announcement is carried out by transmission of a DSC urgency call on the DSC distress calling channel(2,187.5 kHz on MF, channel 70 on VHF).

The urgency message is transmitted on the distress traffic channel (2,182 kHz on MF, channel 16 on VHF).

The DSC urgency call may be addressed to all stations or to a specific station. The frequency on which the urgency message will be transmitted shall be included in the DSC urgency call.

The transmission of an urgency message is thus carried out as follows:

Announcement

- tune the transmitter to the DSC distress calling channel (2,187.5 kHz on MF, channel 70 on VHF);
- key in or select on the DSC equipment keyboard:
    - All Ships Call or the 9-digit identity of the specific station,
    - the category of the call (urgency),
    - the frequency or channel on which the urgency message will be transmitted.
    - the type of communication in which the urgency message

will be given (radiotelephony),

in accordance with the DSC equipment manufacturer's instructions;

- transmit the DSC urgency call.

Transmission of the urgency message:

- tune the transmitter to the frequency or channel indicated in the DSC urgency call;
- transmit the urgency message as follows:
  - "PAN PAN", repeated 3 times,
  - "ALL STATIONS" or called station. repeated 3 times,
  - "this is"
  - the 9-digit identity and the call sign or other identification of own ship,
  - the text of the urgency message.

2.2 Reception of an urgency message

Ships receiving a DSC urgency call announcing an urgency message addressed to all ships shall NOT acknowledge the receipt of the DSC call, but should tune the radiotelephony receiver to the frequency indicated in the call and listen to the urgency message.

3. Safety

3.1 Transmission of safety messages

Transmission of safety messages shall be carried out in two steps:

- announcement of the safety message,
- transmission of the safety message.

The announcement is carried out by transmission of a DSC safety call on the DSC distress calling channel(2,187.5 kHz on MF, channel 70 on VHF).

The safety message is normally transmitted on the distress and safety traffic channel in the same band in which the DSC call was sent, i.e. 2,182 kHz on MF, channel 16 on VHF.

The DSC safety call may be addressed to all ships, all ships in a specific geographical area or to a specific station.

The frequency on which the safety message will be transmitted shall be included in the DSC call.

The transmission of a safety message is thus carried out as follows:

Announcement:

- tune the transmitter to the DSC distress calling channel (2,187.5 kHz on MF, channel 70 on VHF);
- select the appropriate calling format on the DSC equipment (all ships, area call or individual call);
- key in or select on the DSC equipment keyboard:
  - specific area or 9-digit identity of specific station, if appropriate,
  - the category of the call (safety),
  - the frequency or channel on which the safety message will be transmitted
  - the type of communication in which the safety message

will be given (radiotelephony),

in accordance with the DSC equipment manufacturer's instructions;

- transmit the DSC safety call.

Transmission of the safety message :

- tune the transmitter to the frequency or channel indicated in the DSC safety call ;
- transmit the safety message as follows:
  - "SECURITE", repeated 3 times,
  - "ALL STATIONS" or called station, repeated 3 times,
  - "this is",
  - the 9-digit identity and the call sign or other identification of own ship,
  - the text of the safety message.

3.2 Reception of a safety message

Ships receiving a DSC safety call announcing a safety message addressed to all ships shall NOT acknowledge the receipt of the DSC safety call, but should tune the radiotelephony receiver to the frequency indicated in the call and listen to the safety message.

4. Public correspondence

4.1 DSC channels for public correspondence

4.1.1. VHF

The VHF DSC channel 70 is used for DSC for distress and safety purposes as well as for DSC for public correspondence.

4.1.2 MF

International and national DSC channels separate from the DSC distress and safety calling channel 2,187.5 kHz are used for digital selective-calling on MF for public correspondence.

Ships calling a coast station by DSC on MF for public correspondence should preferably use the coast station's national DSC channel.

The international DSC channel for public correspondence may as a general rule be used between ships and coast stations of different nationality. The ship's transmitting frequency is 2,189.5 kHz, and the receiving frequency is 2,177 kHz.

The frequency 2,177 kHz is also used for digital selective-calling between ships for general communication

4.2 Transmission of a DSC call for public correspondence to a coast station or another ship

A DSC call for public correspondence to a coast station or another ship is transmitted as follows:

- tune the transmitter to the relevant DSC channel;
- select the format for calling a specific station on the DSC equipment;
- key in or select on the DSC equipment keyboard:
  - the 9-digit identity of the station to be called,
  - the category of the call (routine),

  - the type of the subsequent communication (normally radiotelephony),
  - a proposed working channel if calling another ship. A proposal for a working channel should NOT be included in calls to a coast station; the coast station will in its DSC acknowledgement indicate a vacant working channel,,in accordance with the DSC equipment manufacturer's instructions;
- transmit the DSC call.

4.3 Repeating a call

A DSC call for public correspondence may be repeated on the same or another DSC channel, if no acknowledgement is received within 5 min.

Further call attempts should be delayed at least 15 min, if acknowledgement is still not received

4.4 Acknowledgement of a received call and preparation for reception of the traffic

On receipt of a DSC call from a coast station or another ship, a DSC acknowledgement is transmitted as follows:

- tune the transmitter to the transmit frequency of the DSC channel on which the call was received,
- select the acknowledgement format on the DSC equipment,
- transmit an acknowledgement indicating whether the ship is able to communicate as proposed in the call (type of communication and working frequency),
- if able to communicate as indicated, tune the transmitter and the radiotelephony receiver to the indicate working

channel and prepare to receive the traffic.

4.5 Reception of acknowledgement and further actions

When receiving an acknowledgement indicating that the called station is able to receive the traffic, prepare to transmit the traffic as follows:

- tune the transmitter and receiver to the indicated working channel ;
- commence the communication on the working channel by:
  - the 9-digit identity or call sign or other identification of the called station
  - "this is"
  - the 9-digit identity or call sign or other identification of own ship.

It will normally rest with the ship to call again a little later in case the acknowledgement from the coast station indicates that the coast station is not able to receive the traffic immediately.

In case the ship, in response to a call to another ship, receives an acknowledgement indicating that the other ship is not able to receive the traffic immediately, it will normally rest with the called ship to transmit a call to the calling ship when ready to receive the traffic.

부록 4

[4] 무선전화 통신용 약어 및 표준 어귀

1. 약어 (Abbreviations)

국 문	영 문
조난	MAYDAY
긴급	PAN PAN
경보	SECURITY
의료	MEDICAL
각국 또는 각선	HELLO ALL SHIPS/STATIONS
여기는	THIS IS (DELTA ECHO)
말씀하시오(계속하시오)	OVER (GO AHEAD)
정정	CORRECTION
반복	REPEAT, READ BACK, SAY AGAIN
수신하였습니다	RECEIVED(ROMEO), ROGER
명료도	READABILITY
강도	STRENGTH
시험 중	TEST, SIGNAL CHECK
하나 둘 셋 넷 …	ONE TWO THREE FOUR …
…의 뒤 전부	ALL AFTER …
…의 앞 전부	ALL BEFORE …
…의 뒷 단어	WORD AFTER …
…의 앞 단어	WORD BEFORE …
귀국에 보낼 통신이 있습니다	MESSAGE FOR YOU
통신이 더 있습니다	MORE TO FOLLOW
직무 끝	OFF WORK
대기 중	STAND BY
통신 종료	OUT

2. VHF 무선통신의 표준어구(Standard Phrases)

(1) 통신 연락 설정 및 유지(Making and maintaining contact)

1) … 을(로) 호출하고 있음:	Calling …
2) 당국의 감도는?:	How do you read?
3) 귀국의 감도는…(1-5)임:	I read you …(1-5)
4) 송화 끝, 말씀하시오:	Over
5) 통신 종료:	Out
6) VHF 채널…로 대기하시오:	Stand by on VHF channel …
7) VHF 채널…로 대기중임:	Standing by on VHF channel…
8) …분 동안 기다려 주시오:	Wait … minutes

(2) 통화 관제(Conversation controls)

1) 잠시 중단하시오:	Break.
2) 더 이상 통화할 사항 없음:	Nothing more.
3) 수신확인통보 바랍니다:	Please acknowledge.
4) 복창 바랍니다:	Please read back.
5) 복창합니다:	Read back.
6) 양지하였음:	Understood.

(3) 확인(Clarification)

1) 정정:	Correction
2) 착오(발생), 정확한 내용은 다음과 같습니다:	Mistake
3) 약어 대신 완전한 송화바랍니다:	Please speak in full.
4) 천천히 송화하여 주시오:	Please speak slowly.
5) 단어의 철자를 통보해 주시오:	Please spell …
6) 단어의 철자는 다음과 같습니다:	I spell …
7) SEASPEAK를 사용해 주시오:	Please use SEASPEAK.
8) 반복해 주시오:	Say again.
9) 반복합니다:	I repeat.(I say again)

* Mistake : There is a mistake in your last transmission.
The information should be …

(4) 안내 방송(Announcements)

1) 최종 호출:	Final call
2) 귀국에 전송할 통신문이 있습니다:	Message for you.
3) 귀국의 통신을 전송하시오;	Pass your message.

(5) 채널 변경(Channel switching)

1) VHF 채널…로 나오시오:	On VHF channel …
2) VHF 채널…로 변경하시오:	Switch to VHF channel …
3) VHF채널…의 사용을 동의합니다:	Agree VHF channel …
4) 이용 가능한 VHF채널은…입니다:	VHF channel … available
5) VHF채널…은 이용할 수 없습니다;	VHF channel … unable.
6) 어느 VHF 채널입니까?:	Which VHF channel?

3. 통보 부호(Message markers)

(1)질문: QUESTION,	회답	: ANSWER
(2)지시: INSTRUCTION,	지시 수신하였음:	INSTRUCTION-RECEIVED
(3)권고: ADVICE,	권고 수신하였음:	ADVICE-RECEIVED
(4)요청: REQUEST,	요청 수신하였음:	REQUEST-RECEIVED
(5)정보: INFORMATION,	정보 수신하였음:	INFORMATION-RECEIVED
(6)경고: WARNING,	경고 수신하였음:	WARNING-RECEIVED
(7)의도: INTENTION,	의도 수신하였음:	INTENTION-RECEIVED

부록 5

[5] 기상 계급표(Meteorological Scale Table)

1. 너울의 계급 (Swell Scale)

<table>
<tr><th></th><th>명 칭</th><th>Swell의 상태</th></tr>
<tr><td>0</td><td>no swell</td><td>없음</td></tr>
<tr><td>1</td><td>low swell, short or average length</td><td rowspan="2">약한 swell (파고 2m 미만)</td></tr>
<tr><td>2</td><td>low swell, long</td></tr>
<tr><td>3</td><td>moderate swell, short</td><td rowspan="3">다소 높은 swell (파고 2m~4m)</td></tr>
<tr><td>4</td><td>moderate swell, average length</td></tr>
<tr><td>5</td><td>moderate swell, long</td></tr>
<tr><td>6</td><td>heavy swell, short</td><td rowspan="3">높은 swell (파고 4m 이상)</td></tr>
<tr><td>7</td><td>heavy swell, average length</td></tr>
<tr><td>8</td><td>heavy swell, long</td></tr>
<tr><td>9</td><td>confused swell</td><td>두 방향에서 swell이 오고
해상은 혼란함</td></tr>
</table>

* short는 파장 100m미만(주기 8.0초 이하)
* long은 파장 200m이상(주기11.4초 이상)
* average는 파장 100m~200m(주기 8.1초~11.3초)

2. 파랑의 계급 (Wave Scale, State of Sea)

계 급	명 칭	파 고 (m)
0	calm(glassy)	0
1	calm(rippled)	0 ~ 1/10
2	smooth(wavelets)	1/10 ~ 1/2
3	slight	1/2 ~ 1¼
4	moderate	1¼ ~ 2½
5	rough	2½ ~ 4
6	very rough	4 ~ 6
7	high	6 ~ 9
8	very high	9 ~ 14
9	phenomenal	14 ~

3. 시정표 (Visibility Scale)

계급	명 칭	시인 거리
0	Dense fog	약 50m 이하 (~50m)
1	Thick fog	약 1 cable 이하 (50m~200m)
2	Fog	약 2 cable 이하 (200~500m)
3	Moderate fog	1/2 mile 이하 (500m~1km)
4	Thin fog or mist	1 mile 이하 (1km~2km)
5	Poor visibility	2 mile 이하 (2km~4km)
6	Moderate visibility	5 mile 이하 (4km~10km)
7	Good visibility	10 mile 이하 (10km~20km)
8	Very good visibility	30 mile 이하 (20km~50km)
9	Exceptional visibility	30 mile 이상 (50km 이상)

4. 풍력 계급표(Beaufort Scale of Wind Force)

Beaufort No.	Knots (MPH)	M/Sec.	Description
0	Less than 1	0~0.3	Calm
1	1~4	0.3~1.6	Light air
2	4~7	1.6~3.4	Light breeze
3	7~11	3.4~5.5	Gentle breeze
4	11~17	5.5~8.0	Moderate breeze
5	17~22	8.0~10.8	Fresh breeze
6	22~28	10.8~13.9	Strong breeze
7	28~34	13.9~17.2	Near gale
8	34~41	17.2~20.8	Gale
9	41~48	20.8~24.5	Strong gale
10	48~56	24.5~28.5	Storm
11	56~64	28.5~32.7	Violent Storm
12	64~	32.7~	Hurricane

부록 6

[6] GENERAL PROVISIONS ON SHIPS' ROUTEING

1. OBJECTIVES

1.1 The purpose of ships' routeing is to improve the safety of navigation in converging areas and in areas where the density of traffic is great or where freedom of movement of shipping is inhibited by restricted sea-room, the existence of obstructions to navigation, limited depths or unfavourable meteorological conditions.

▷Ships' routeing may also be used for the purpose of preventing or reducing the risk of pollution or other damage to the marine environment caused by ships colliding or grounding in or near environmentally sensitive areas.◁

1.2 The precise objectives of any routeing system will depend upon the particular hazardous circumstances which it is intended to alleviate, but may include some or all of the following:

.1 the separation of opposing streams of traffic so as to reduce the incidence of head-on encounters;

.2 the reduction of dangers of collision between crossing traffic and shipping in established traffic lanes;

.3 the simplification of the patterns of traffic flow in converging areas;

.4 the organization of safe traffic flow in areas of concentrated off-shore exploration or exploitation;

.5 the organization of traffic flow in or around areas where navigation by all ships or by certain classes of ship is dangerous or undesirable;

.6 the reduction of risk of grounding by providing special guidance to vessels in areas where water depths are uncertain or critical:

.7 the guidance of traffic clear of fishing grounds or the organization of traffic through fishing grounds.

2. DEFINITIONS

2.1 The following terms are used in connection with matters related to ships' routeing:

.1 Routeing system

Any system of one or more routes or routeing measures aimed at reducing the risk of casualties; it includes traffic separation schemes, two-way routes, recommended tracks, areas to be avoided, inshore traffic zones, roundabouts, precautionary areas and deep-water routes.

.2 Traffic separation scheme

A routeing measure aimed at the separation of opposing streams of traffic by appropriate means and by the establishment of traffic lanes.

.3 Separation zone or line

A zone or line separating the traffic lanes in which ships are proceeding in opposite or nearly opposite directions; or separating a traffic lane from the adjacent sea area; or separating traffic lanes designated for particular

classes of ship proceeding in the same direction.

.4 Traffic lane

An area within defined limits in which one-way traffic is established. Natural obstacles, including those forming separation zones, may constitute a boundary.

.5 Roundabout

A routeing measure comprising a separation point or circular separation zone and a circular traffic lane within defined limits. Traffic within the roundabout is separated by moving in a counterclockwise direction around the separation point or zone.

.6 Inshore traffic zone

A routeing measure comprising a designated area between the landward boundary of a traffic separation scheme and the adjacent coast, to be used in accordance with the provisions of Rule 10(d), as amended, of the International Regulations for Preventing Collisions at Sea, 1972 (Collision Regulations).

.7 Two-way route

A route within defined limits inside which two-way traffic is established, aimed at providing safe passage of ships through waters where navigation is difficult or dangerous.

.8 Recommended route

A route of undefined width, for the convenience of ships in transit, which is often marked by centreline buoys.

.9 Recommended track

A route which has been specially examined to ensure so far as possible that it is free of dangers and along which ships are advised to navigate.

부록 7

[7] ILLUSTRATIVE DIAGRAM OF SHIP GLOSSARY

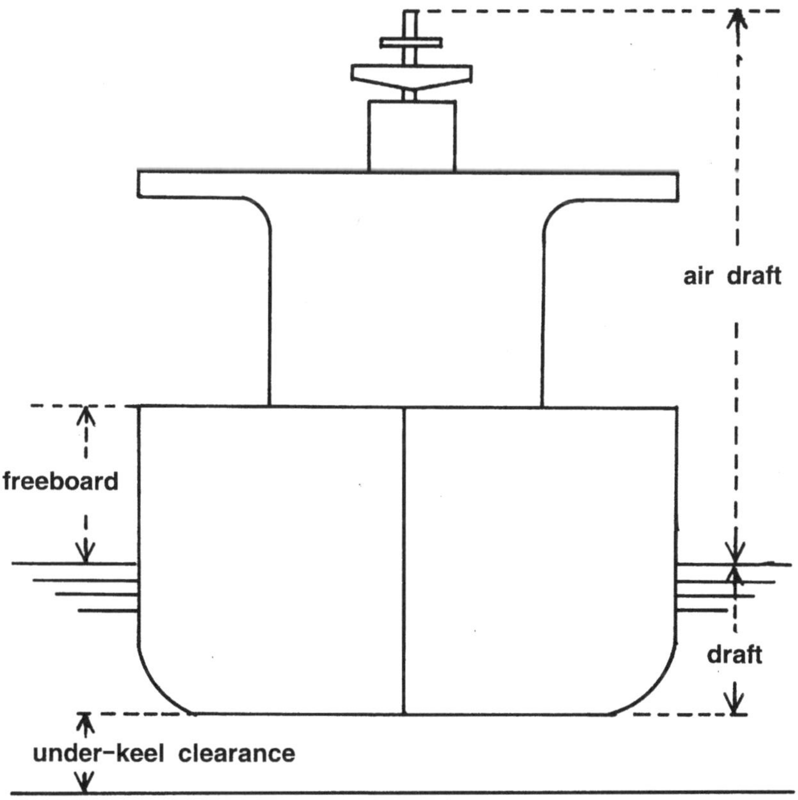

SKETCH 2

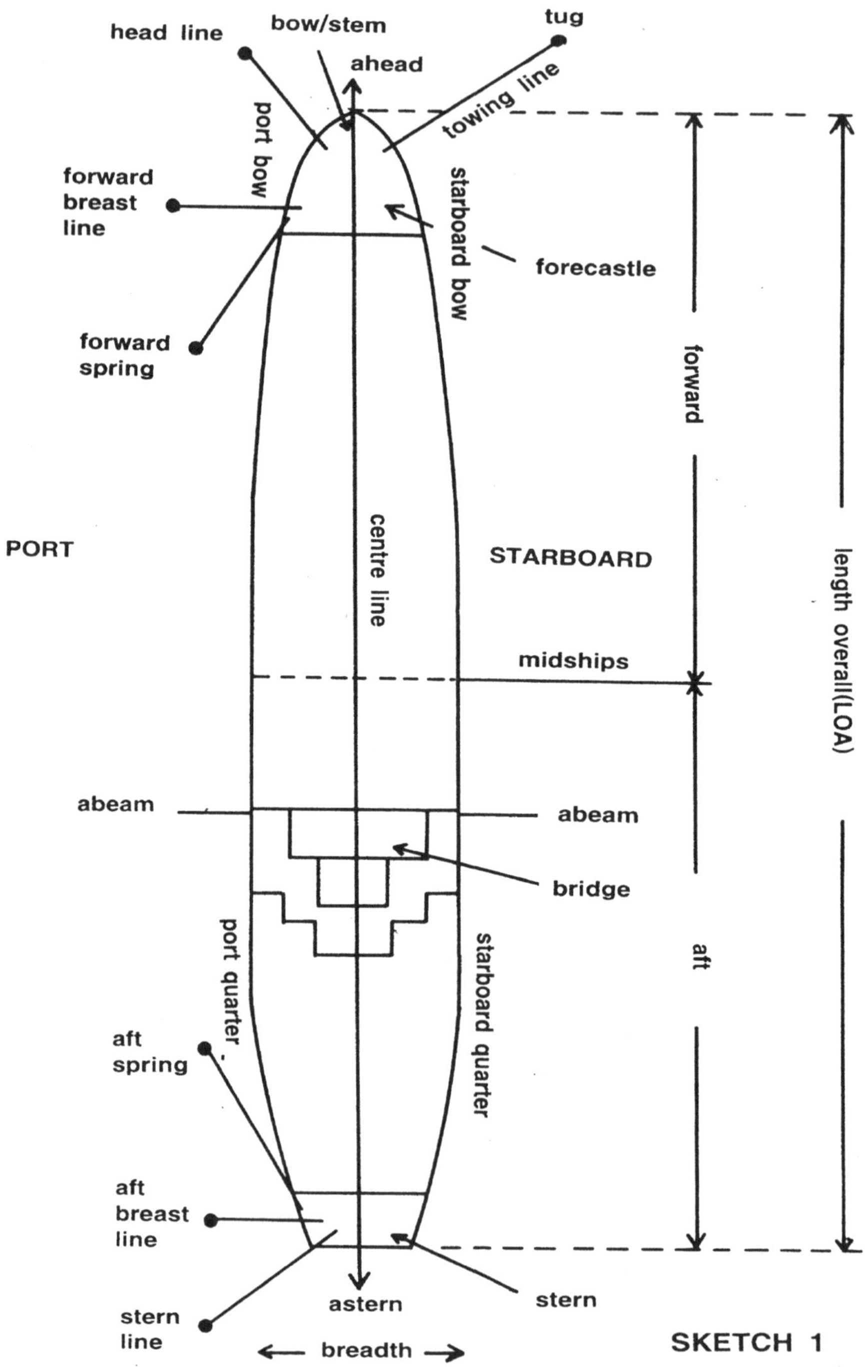
head line
bow/stem
tug
ahead
towing line
port bow
forward
breast
line
starboard bow
forecastle
forward
spring
forward
centre line
PORT
STARBOARD
length overall(LOA)
midships
abeam
abeam
bridge
port quarter
starboard quarter
aft
aft
spring
aft
breast
line
stern
line
astern
stern
breadth

SKETCH 1

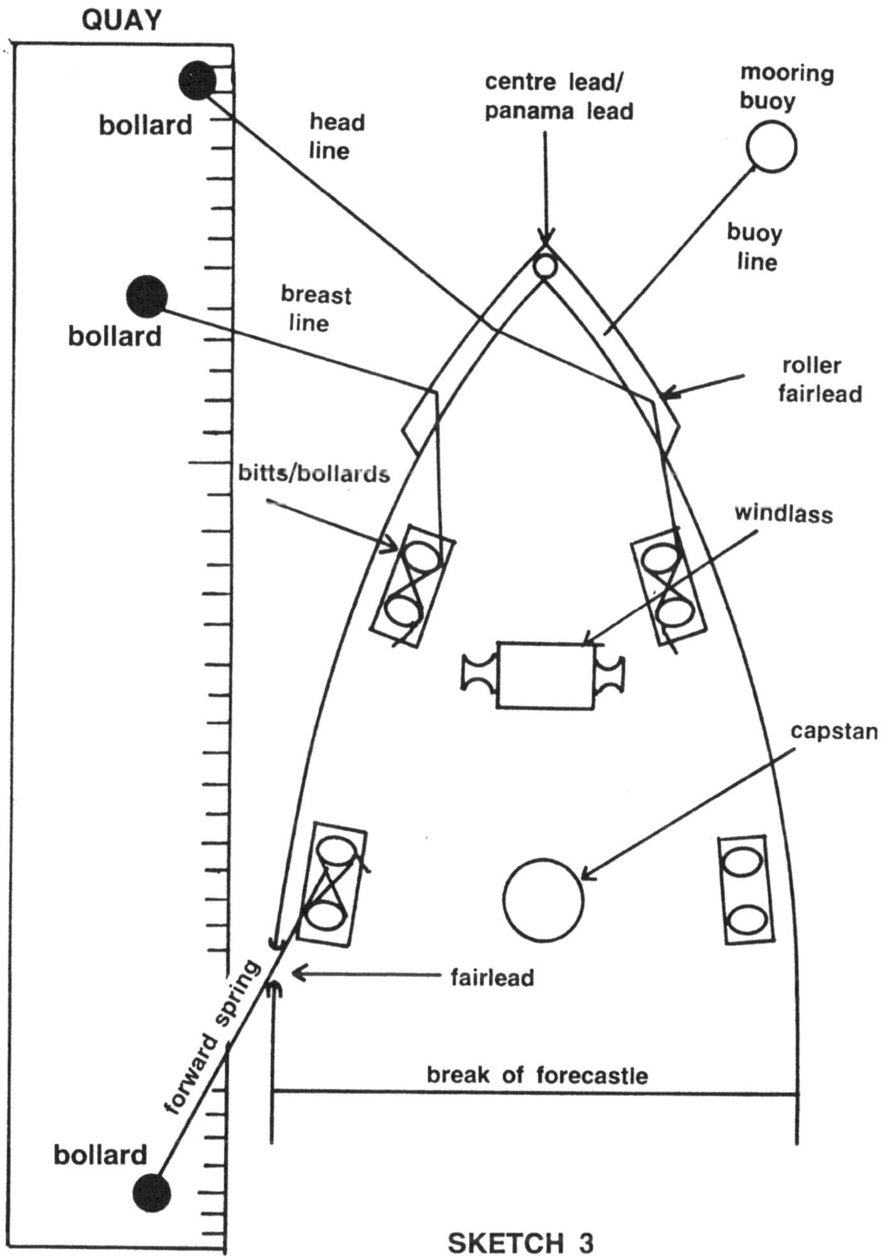
QUAY
bollard
head line
centre lead/ panama lead
mooring buoy
buoy line
bollard
breast line
roller fairlead
bitts/bollards
windlass
capstan
fairlead
forward spring
break of forecastle
bollard

SKETCH 3

參考文獻

o ICS, Bridge Procedure Guide, 海印出版社, 1999

o IMO, GMDSS Handbook, IMO, 1995

o IMO, SHIP'S Routeing (8th Edition), IMO, 2003

o IMO, STCW Convention '95 & STCW Code, IMO 1996
STCW Manila Convention 2010

o IMO, International Code of Signal, IMO, 1987

o IAMSAR, Int'l Aeronautical and Maritime Search and Rescue Manual, Volume III, 해인출판사, 2000

o IMO, International Safetynet Manual, IMO, 1994

o IMO, IMO/UNEP Guidelines on Oil Spill Dispersant Application, IMO, 1995

o Nathaniel Bowditch, AMERICAN PRACTICAL NAVIGATOR, National Imagery and Mapping Agency, 2002

o MCA, Marine Guidance Note MGN 324(M+F) - Radio Operational Guidance on the Use of VHFF Radio and AIS at Sea, Maritime and Coastguard Agency, UK, 2006

o Orlando S. Dimailig, et al, Practical Maritime English Conversation, 2008

o P.C van Kluijven, The International Maritime Language Programme, Alk & Heijnen Publishers, 2003

o Peter Trenkner, The IMO Standard Marine Communication Pharases in Dialogues – Students' Course, Unpublished, 2016

o 飯島幸人 等, GMDSS の 解說, 成山堂書店, 1990

o 福地 章, 海洋氣象講座, 成山堂書店, 1997

o 금종수·양동신, 海上搜索救助論, 효성출판사, 1999

o 김우숙, 레이더항법, 海文出版社, 1985

o 문채식, 海上安全論, 海印出版社, 1996

o 박진수·박영수·이형기, (最新) 海上交通工學, 다솜출판사, 2005

o 박진수·이형기·정재용, (증보) 해사안전관리, 다솜출판사, 2007

o 신철호·김우숙, 전자항해학, 효성출판사, 2001

o 양시권·김순갑, 船舶積貨, 韓國海洋大學校 海事圖書出版部, 1990
o 오정철, 船舶整備論, 세진사, 1991
o 윤점동, 船舶操縱의 理論과 實務, 세종출판사, 1993
o 윤점동, 國際海上衝突豫防規則 및 관련된 國內 法規解說, 세종출판사, 1994
o 이재우, 標準海事航海英語, 海印出版社, 1996
o 최충연, 電波關係法規, 木浦海洋大學校 圖書出版部, 1994
o 韓國船級, 1974년 海上人命安全協約, 海印出版社, 1998
o 한국해양수산연수원, GMDSS 통신운용, 세종문화사, 2014
o 한국해양수산연수원, GMDSS-GOC 교재, 동원문화사 *i*, 2016

해광
출판
해기

IMO 표준해사통신영어

IMO Standard Marine Communication Phrases

2007년 8월 31일 초판 발행 정가 27,000원
2010년 8월 31일 개정1판 발행
2014년 11월 20일 개정2판 발행
2016년 11월 11일 개정2판 2쇄 발행
2024년 9월 15일 개정2판 4쇄 발행

저 자 박진수·김현종
편 집 송영미

발 행 인 조준형 http://www.hgbookpub.com
발 행 처 해광출판사 E-mail : munbookcokmr@naver.com
주 소 부산광역시 중구 해관로 41-1 (중앙동2가)
전화번호 (051) 253-0001 팩스 (051) 245-1187
등 록 제 2012-000005호

해광출판사에서는 여러분의 소중한 원고와 함께 할 기회를 기다리고 있습니다. 책으로 엮을 원고나 아이디어가 있으신 분들은 이메일 munbookcokr@naver.com로 책에 대한 간단한 개요와 원고 전체 또는 일부를 연락처와 함께 보내주십시오.

ISBN 979-11-6881-619-0